Viroids and Viroid-Like Pathogens

Editor

Joseph S. Semancik
Professor
Department of Plant Pathology
University of California, Riverside
Riverside, California

CRC

CRC Press, Inc.
Boca Raton, Florida

Library of Congress Cataloging-in-Publication Data

Viroids and viroid-like pathogens.

Bibliography: p.
Includes index.
1. Viroids. I. Semancik, Joseph S. [DNLM:
1. Viroids. QW 160 V8188]
QR.V57 1987 576'.64 86-21631
ISBN 0-8493-6060-9

Direct all inquiries to CRC Press, Inc., 2000 Corporate Blvd., N.W., Boca Raton, Florida, 33431.

International Standard Book Number 0-8493-6060-9

Library of Congress Card Number 86-21631
Printed in the United States

PREFACE

At the outset I would like to presume the position of at least the caring, if not always humble, ''midwife'' at the birth of viroid research. Without further claims of any closer lineage, that position in itself has permitted me to bear direct witness to not only the pains and anxieties normally associated with such creative activities but also the fleeting yet sublime exhilaration that is born of such wondrous moments. It is my view that viroids because of their humble beginnings in the Plant Kingdom have not as yet reached a rightful position of full acceptance in the biological world. This is in part due to absence of a direct bloodline to relatives affecting members of the Animal Kingdom. From the perspective of an advocate of the existence of great unifying principles of biology, I retain the vision that this linkage will yet be made. For this reason, I have coerced Prof. R. Marsh to contribute a brief chapter on the most analogous agents in animal systems for which research data are available, namely the scrapie disease.

Once the viroids were recognized as discrete biochemical entities in the early 1970s, the science of a biological phenomenon seemingly advanced to a period mirroring the energies of adolescence. This period of viroid research development was reflective of some of the tribulations resident to such a stage in which hypotheses, propositions, and even personal intuitions were presented in spirited dialogue; and, as with teenage romance, it became too serious, too soon, only to fade too quickly.

But the undeniable forces of maturation to which we are all inevitably held captive had been primed. Substantive physical measurements established a legitimate position for viroids at least as an interesting class of unique RNAs from which structural and conformational information might be derived. We are presently heir to that treasure of information as presented in the chapter contributred by Dr. Paul Keese and Prof. Symons.

And yet without the further investment of these resources, we would be soon left with but a dwindling fortune to will to the succeeding generation of true believers in the basic importance of viroids in the grand scheme of biology. And so when I was approached with the siren and at the same time ominous proposition by CRC Press for this volume, I immediately enlisted the active participation of Robert, Hugh, and Stephen to comprise yet another set of ''four evangelists.'' In this way it was hoped that a truth of purpose might emerge from a consensus position. My only directive was to encourage not simply another review, but a personal perspective of interpretive data and a willingness to offer explanation and even speculation on the significance of viroid synthesis and biological interactions. Viroid replication viewed as an RNA processing event has been contributed by Prof. Robertson and Dr. Andrea Branch. We have chosen to package the responses of plants to viroids with the broader agricultural insights of Dr. Garnsey and Dr. Randles, and from a more cellular position, where it has been my pleasure to interact once again with my unique friend Prof. Conejero.

If we had hoped for one unifying theme among these chapters, it would be contained in the belief that the future of viroid research resides in the basic biochemical interactions with the plant cell that result in processes which alter growth and development, some of which may be labelled as pathological reactions. And that these expressions, which we hope are made more relevant by this volume, offer even more exciting challenges and resident rewards than even the initial conception of viroids as entities would be our wish.

We would simply entreat the reader along with the fair Ophelia to not only ''let all 'our' sins be remembered'' but also forgiven.

2 May in 1986
Riverside, California

THE EDITOR

Joseph S. Semancik, Ph.D., is Professor of Plant Pathology and a member of the Cell Interaction Research Group at the University of California, Riverside.

Professor Semancik received the A.B. degree (cum laude) in biology in 1960 from Western Reserve University, Cleveland, Ohio. He received the M.S. (1962) and Ph.D. (1964) from Purdue University, W. Lafayette, Indiana, working under the supervision of Professor J. B. Bancroft, in part, as a NIH predoctoral fellow. In 1964, he joined the faculty of the University of California as Assistant Professor of Plant Pathology and, except for the period of 1969-71, when he held the position of Associate Professor at the University of Nebraska-Lincoln, he has continued his affiliation with the University of California, Riverside.

The research interests of Professor Semancik evolved from his early biochemical and biophysical studies of the basis for electrophoretic and centrifugal heterogeneity of plant viruses to detection and properties of free-RNA forms of plant virus infection. His current program is directed exclusively to viroids, with particular emphasis on detection, synthesis, and viroid-cell interactions. Since 1968, he, along with students and colleagues in his laboratory, have contributed 45 published papers specifically to the field of viroid research.

Professor Semancik received the Spengler Award for Excellence in Botany from Western Reserve University (1960). In 1977-78, he was named a Guggenheim Fellow and a NATO Senior Scientist in support of studies in the Department of Biochemistry, State University, Leiden, The Netherlands. In 1984 he participated in a NATO Collaborative Research Project in Valencia, Spain, investigating the viroids of the region.

The American Phytopathological Society designated Professor Semancik a Fellow in 1979. In 1975, Professor Semancik shared with Dr. T. O. Diener the Alexander von Humboldt Award for "research in the recognition and establishment of viroids, a hitherto unknown group of pathogens and for their biological and chemical characterization."

CONTRIBUTORS

Andrea D. Branch
The Rockefeller University
New York, New York

V. Conejero-Tomas
Escuela Tecnica Superior de Ingenieros Agronomos
Universidad Politecnica
Valencia, Spain

Stephen M. Garnsey
Agricultural Research Service
U.S. Department of Agriculture
Orlando, Florida

P. Keese
Adelaide University Centre for Gene Technology
University of Adelaide
Adelaide, South Australia

R. F. Marsh
Department of Veterinary Science
University of Wisconsin-Madison
Madison, Wisconsin

J. W. Randles
Waite Agricultural Research Institute
University of Adelaide
Adelaide, South Australia

Hugh D. Robertson
The Rockefeller University
New York, New York

Robert H. Symons
Adelaide University Centre for Gene Technology
University of Adelaide
Adelaide, South Australia

TABLE OF CONTENTS

Chapter 1

THE STRUCTURE OF VIROIDS AND VIRUSOIDS

P. Keese and R. H. Symons

TABLE OF CONTENTS

I. INTRODUCTION

A. Viroids

Viroids (defined in the Glossary at the end of this chapter and listed in Table 1) are the smallest known pathogens of flowering plants. Their history is brief but dramatic, both in appreciation of their ecomonic impact and in the growth of knowledge concerning their unique characteristics. Viroid-induced diseases were first recognized only recently with the earliest reports dating back to 1922 for potato spindle tuber disease (see References 19 and 22 for a detailed account of the history of viroid diseases). Nevertheless, viroids have been responsible for several economically serious diseases such as chrysanthemum stunt diseases in the U.S. from 1945 to 1947[23] and cadang cadang disease of coconuts which has resulted in the death of more than 30 million palms in the Philippines since its discovery in 1930.[24] Cadang cadang disease remains uncontrolled and continues to spread (see Chapter 4).

Table 1
LIST OF VIROIDS

Viroid[a]	Abbreviation	No. of nucleotides	Ref.
Avocado sunblotch viroid	ASBV	247	3,4
Chrysanthemum stunt viroid	CSV	354,356	5,6
Citron variable viroid	CVaV	N.D.[e]	7
Citrus exocortis viroid	CEV	370—375	6,8-10
Coconut cadang cadang viroid[b]	CCCV	246,247	11
Coconut tianangaja viroid	CTiV	N.D.[e]	20,21
Columnea viroid	CV	N.D.[e]	12
Cucumber pale fruit viroid[c]	CPFV	303	13
Hop stunt viroid	HSV	297	14
Potato spindle tuber viroid	PSTV	359	15,16
Tomato apical stunt viroid[d]	TASV	360	17
Tomato planta macho viroid	TPMV	360	17

[a] The disease agent of Burdock stunt has been considered to share some affinities with viroids.[18] The disease agent of chrysanthemum chlorotic mottle disease has often been listed as a viroid.[19] However, since these disease agents have never been isolated, they are not included in this table.

[b] CCCV infections produce four major RNA components, all derived from the infectious monomeric small form (D-O of 246 or 247 nucleotides). These include monomers and dimers of both the D-O form and any of a set of larger forms (D-41, D-50, or D-55) which contain a duplication involving 41, 50, or 55 nucleotides. The causative agent (CTiV) of another disease of coconuts, tinangaja[20] is approximately the same size as, and shows partial sequence homology with, CCCV.[20,21]

[c] CPFV is a sequence variant of HSV since the two viroids share 95% sequence homology.[13,14]

[d] In early reports, TASV was referred to as tomato bunchy top viroid.[19]

[e] N.D., not determined.

Although viroid diseases were initially linked to viruses due to the virus-like symptoms (see Chapter 3) and the small size of the causative agent, the distinctive features of potato spindle tuber "virus" were made apparent by Diener in 1971,[22] leading to the term viroid. Similar properties were reported for the citrus exocortis and chrysanthemum stunt agents.[25-28] Since then, eight additional viroids have been recognized (Table 1). Furthermore, there has been a rapid accumulation of information regarding viroid structure, replicaton, and pathogenesis.

This chapter will present the most recent findings and postulations which attempt to relate directly viroid structure to function. For a more general introduction to viroids, the reader is referred to the excellent reviews of Riesner et al.,[29] Sänger,[30] Diener,[31] Riesner and Gross,[32] and the book by Diener.[19]

Summarizing the present state of knowledge, viroids can be distinguished from viruses by:

1. **Lack of mRNA activity.** This has two important consequences. First, there is no viroid-coded protein coat; a virus-coded protein coat has normally been associated with virus survival and spread. Second, viroids rely completely on host factors for their replication in contrast to viruses which have been shown to encode a viral-specific polymerase in all cases where definitive results have been obtained.

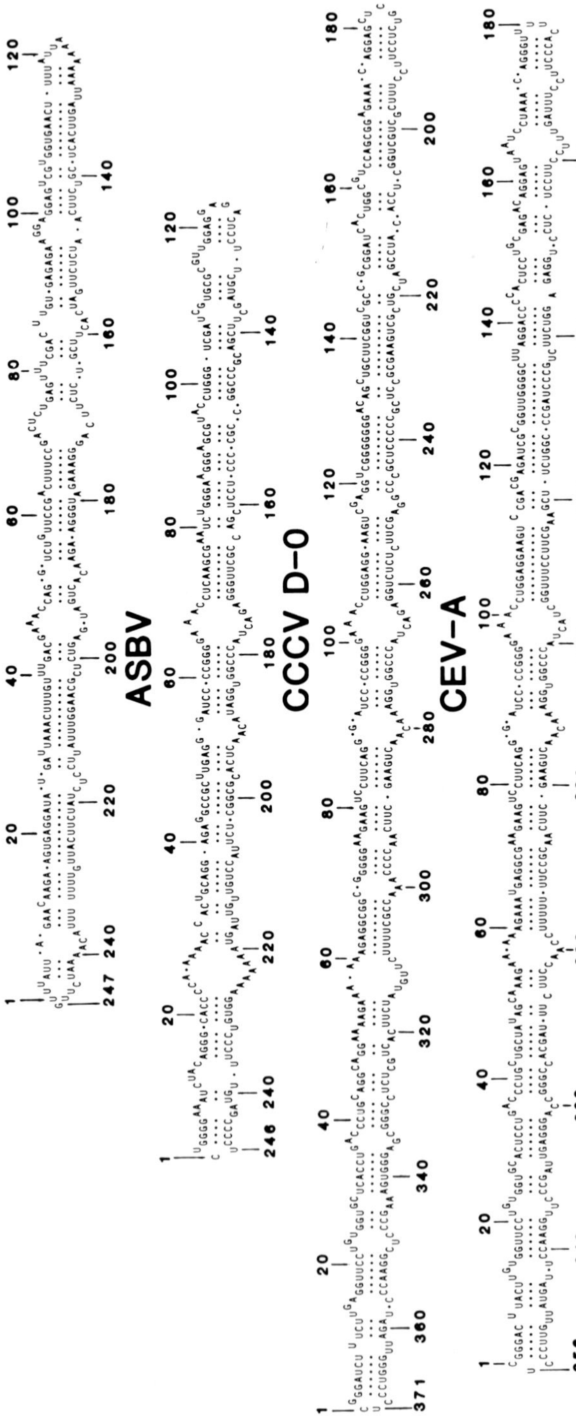
ASBV
CCCV D-O
CEV-A
CSV

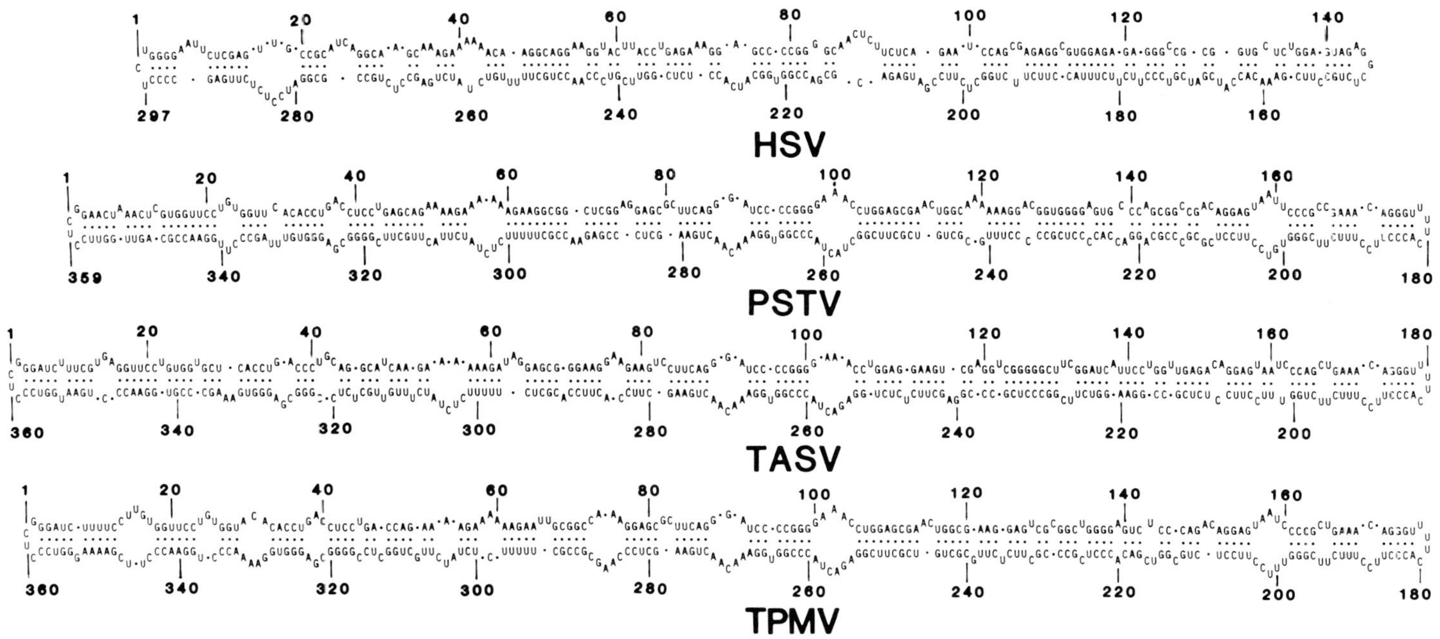

FIGURE 1. Sequences and proposed secondary structures of ASBV,[4] CCCV D-O (the basic 246 nucleotide species),[11] CEV-A,[8] CSV,[5] HSV,[14] PSTV,[15] TASV,[17] and TPMV.[17] CPFV[13] (not given) is a sequence variant of HSV (Table 1).

Table 2
LIST OF VIRUSOIDS

Virusoid	Abbreviation	No. of nucleotides	Ref.
Lucerne transient streak virus RNA 2	vLSTV	324	39,45
Solanum nodiflorum mottle virus RNA 2	vSNMV[a]	378	40,43,44
Subterranean clover mottle virus RNA 2	v(388)SCMoV[b]	388	41,48
Subterranean clover mottle virus RNA 2′	v(332)SCMoV[b]	332	41,48
Velvet tobacco mottle virus RNA 2	vVTMoV[a]	366,367	40,42,44

[a] vSNMV is a sequence variant of vVTMoV since they share 93% sequence homology.[40]
[b] Natural isolates of SCMoV contain either one or two virusoids.[48]

2. **Small size.** With 246 nucleotides, CCCV D-O (Table 1, Figure 1) is less than one tenth the size of the genomes of the smallest known viruses and bacteriophages such as maize streak virus, a single-stranded (ss) DNA virus (2681 nucleotides,[33] 2687 nucleotides[34]), or a ssRNA bacteriophage, MS2 (3569 nucleotides[35]).
3. **Increased yields and symptoms at elevated temperatures.** At temperatures above 20°C and at least up to 35°C, the rate of viroid replication increases and hastens the onset of symptoms.[30,36,37] This is in contrast to plant viruses where high temperatures have been used as a method of curing seeds, bulbs, and cuttings of virus.

With the complete sequence of eight viroids (Figure 1) and more than 35 sequence variants (see Glossary for definition), together with a wealth of structural data,[29,38] viroids have become one of the best structurally characterized groups of RNA molecules. The correlation of structure to the biology of viroids, however, remains poorly understood. Despite this lack of understanding, the total dependence of viroids on host functions, utilizing only sequence and structural signals, indicates that plants can replicate RNA molecules and this opens up the potential for biochemical control. It also suggests the ability of RNA sequences to act directly as control elements in the entire replication cycle and pathogenic pathway involving host range and symptom expression.

B. Virusoids

In addition to viroids, there is a second group of low molecular weight ssRNAs associated with plant diseases; these are the plant virus satellite RNAs (see Glossary; References 1 and 2). Although some of these RNAs also do not show mRNA activity, they differ from viroids in two significant aspects: (1) they are dependent on a helper virus for their replication; (2) they are encapsidated by either viral-coded or satellite RNA-coded coat protein. Amongst the satellite RNAs, the virusoids (see Glossary) show most similarity to viroids in being covalently closed circular RNA molecules in the same size range, varying from 324 to 388 nucleotides (Table 2, Figure 2).

The first report of this new group of circular RNAs was in 1981 with the discovery of velvet tobacco mottle virus (VTMoV[42]) in the desert regions of northern South Australia. When total virion RNA was fractionated by polyacrylamide gel electrophoresis, two RNA components were found, one being a single-stranded linear molecule of M_r 1.5×10^6 and the other a low molecular weight RNA subsequently shown to be single-stranded and circular with 355 or 356 nucleotides.[40,42] Re-investigation of two previously described viruses, solanum nodiflorum mottle virus (SNMV[40,43,44]) and lucerne transient streak virus (LTSV[39,45])

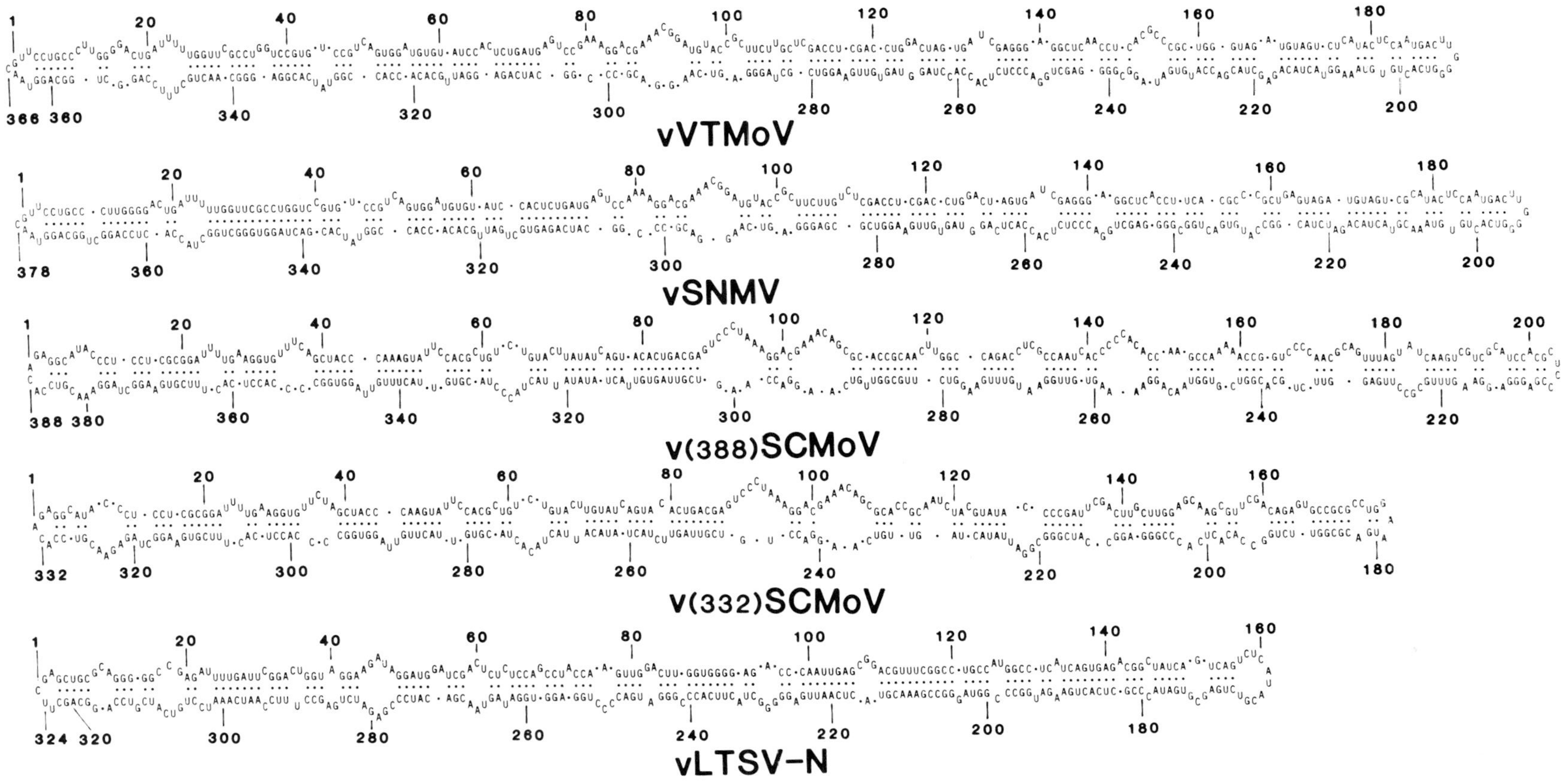

FIGURE 2. Sequences and proposed secondary structures of virusoids. References and further details are in Table 2.

showed that each contained a virusoid. The failure to detect virusoids in the previously reported SNMV and LTSV was probably due to their small size and their confusion with breakdown products of the genomic RNA.[46,47] The most recently reported virusoids were found associated with a new virus infecting subterranean clover *(Trifolium subterraneum)*, subterranean clover mottle virus (SCMoV[48]).

Of these four virusoid-containing viruses, only LTSV has been reported outside Australia and New Zealand.[49] They are considered to belong to the Sobemovirus group of plant viruses with a unipartite ssRNA genome of about 4400 nucleotides with a small viral protein (Vpg) attached at the 5′-terminus.[50]

Virusoids were originally considered to be an essential component for infection, suggesting that they acted in a bipartite relationship with the large genomic RNA rather than as a satellite RNA.[51] Subsequent data have shown that, at least in the cases of SNMV, LTSV, and VTMoV, the large RNA component can replicate independently,[52-54] whereas virusoids are not infectious,[51-53] indicating that the virusoids of these three viruses are satellite RNAs.

Recently, the satellite RNA of tobacco ringspot virus (sTRSV; Figure 3) has been shown to have high sequence homology with a 50 nucleotide sequence in the virusoids of SNMV and VTMoV[55-57] which was somewhat surprising since sTRSV is associated with a different group of plant viruses, the Neopviruses.[50] Furthermore, although the encapsidated forms of sTRSV are linear,[58,59] abundant circular forms exist in vivo.[59,60] Further data are needed to determine if there are functional as well as structural similarities between viroids, virusoids, and sTRSV.

II. PURIFICATION OF VIROIDS AND VIRUSOIDS

Viroids and virusoids exist mostly in infected plants in the circular monomeric plus form with only low or negligible levels of linear monomers. Lower levels of monomeric minus (complementary) forms and of oligomeric plus and minus forms have been detected by hybridization analysis after fractionation of tissue extracts by gel electrophoresis.[61-66] One exception is CCCV where the concentration of dimeric plus CCCV in coconut palms can increase to levels similar to those of the monomer later in infection.[67,68] Of the other viroids and virusoids, dimers of ASBV, SNMV, and VTMoV can also be detected by staining after gel electrophoresis of plant extracts[69,70] (Figure 4).

A. Purification of Viroids

1. Preparation of Plant Extracts

A commonly used approach has been to prepare a partially purified nucleic acid extract of infected plant tissue and then to purify the circular and linear forms of the viroid by gel electrophoresis. Earlier procedures for the preparation of the partially purified extract[71-73] contained a number of steps and required several days for completion. However, many of these steps can be eliminated; e.g., DNase digestion of the extract to degrade DNA, precipitation of nucleic acids with cetyltrimethyl ammonium bromide (CTAB) chromatography on CF-11 cellulose, and fractionation of nucleic acids into high and low molecular weight fractions in the presence of 2 *M* salt at 0°C.

During the preparation of the partially purified extract, phenol is still the best deproteinizing agent and it also has the advantage of removing considerable colored material. In order to circumvent the use of large volumes of phenol when 500 g quantities or more of plant material were used, we developed a two-step extraction procedure[69,74,75] which has proven very effective for ASBV, CSV, CEV, and PSTV. In the first step, plant material was homogenized in the presence of SDS a range of compounds to inhibit color development

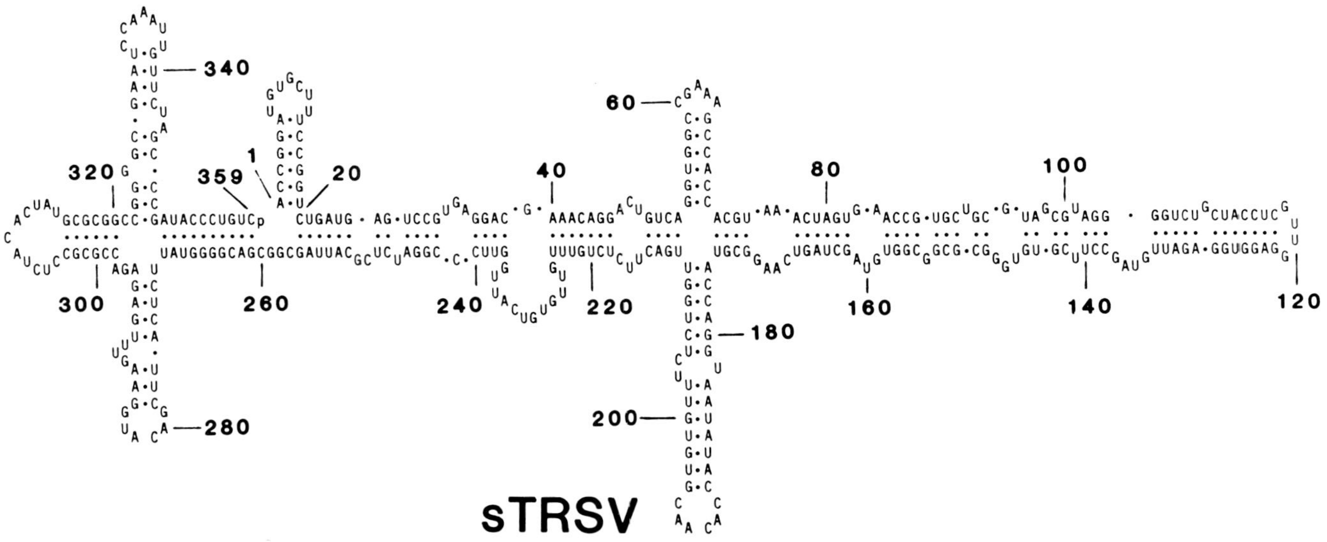

FIGURE 3. Sequence and proposed secondary structure of satellite RNA of tobacco ringspot virus (sTRSV).[55]

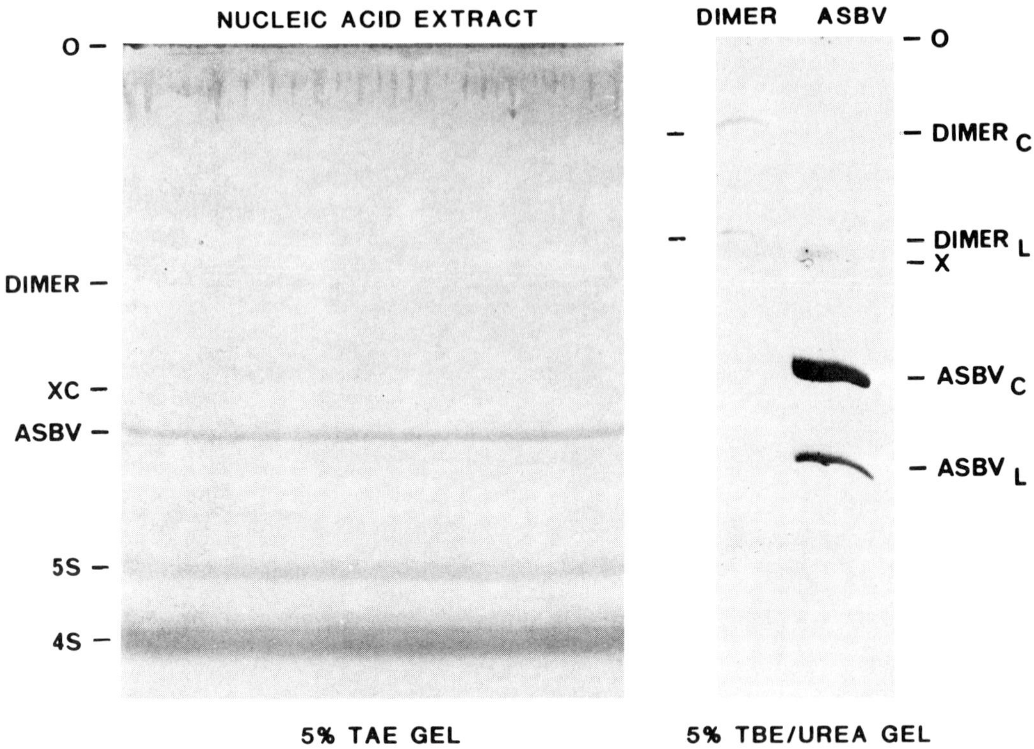

FIGURE 4. Purification of ASBV from extracts of infected avocado leaves.[69] (A) Partially purified nucleic acid extract from 35 g leaves was electrophoresed on a 16 × 16 × 0.3 cm 5% polyacrylamide gel in 40 m*M* Tris-acetate, 20 m*M* sodium acetate, 2 m*M* EDTA, pH 7.5, at 25 mA for 16 hr. The gel was stained in 0.02% toluidine blue and destained in water. The positions of the dimer ASBV, xylene cyanol FF marker dye (xc), ASBV, and 5S and 4S RNA are shown. (B) Dimer ASBV and ASBV bands were eluted and run on a 90 m*M* Tris-borate, 2 m*M* EDTA, 7*M* urea, 5% polyacrylamide gel (20 × 40 × 0.05 cm) at 20 mA for 1.8 hr. The gel was stained with 10 μg ethidium bromide/ mℓ for 30 min, destained in water, and photographed under UV light. The positions of the circular (C) and linear (L) forms of the dimer ASBV, ASBV, and an unidentified band X are given. The mobility of circular ASBV was 0.4 relative to the xylene cyanol FF marker dye. (From Bruening, G., Gould, A. R., Murphy, P. J., and Symons, R. H., *FEBS Lett.*, 148, 71, 1982. With permission.)

and $MgCl_2$ to inhibit the solubilization of pectins. Addition of NaCl to 0.5 *M* caused the precipitation of proteins as a SDS complex.[76] The nucleic acids in the supernatant were concentrated by ethanol precipitation and further deproteinized with phenol-$CHCl_3$ in a volume about 10% of the original aqueous extract. After dialysis of the aqueous phase and a further ethanol precipitation step, the partially purified extract was ready for fractionation by gel electrophoresis.

A variation of the final dialysis step was found to be very useful where a number of smaller samples of about 25 g of plant material were being extracted.[77] In this method, the aqueous supernatant after phenol-$CHCl_3$ deproteinization was made to 6 *M* LiCl by the addition of an equal volume of 12 *M* LiCl; the final concentration of 6 *M* LiCl was necessary to ensure precipitation of low molecular weight RNAs. After 2 hr at 0°C, the precipitate was collected by centrifugation, redissolved, and the nucleic acids precipitated with ethanol prior to gel electrophoresis. We have also used this procedure extensively for the preparation of nucleic acid extracts of avocado leaves for the indexing of ASBV by the dot-blot procedure using ^{32}P-cDNA probes.[77]

Where only a few grams of plant material were to be extracted, a procedure modified from that of Laulhere and Rozier[78] gave rapid and efficient extraction of viroids and virusoids[61] and has proven effective in the extraction of CCCV from coconut palm leaves. In this procedure, the plant material was thoroughly homogenized with 0.5 *M* sodium acetate, 10 m*M* $MgCl_2$, 20% ethanol, 3% SDS, and an equal volume of water-saturated phenol. After a second phenol:$CHCl_3$ extraction of the aqueous phase, nucleic acids were recovered by ethanol precipitation and then fractionated by gel electrophoresis.

Although leaf material is usually chosen for viroid extraction because of its ease of homogenization and harvesting, other parts of the plant can be used. For example, Colpan et al.[79] found, on a weight basis, concentrations of PSTV in stems and roots of tomato plants which were nearly as high as those in leaves. Schwinghamer[80] has used bark from the main trunk of citrus trees since it contains a higher concentratin of CEV than leaves; extraction was done after pulverizing the bark in liquid nitrogen. In view of the very low level of CEV in infected citrus leaves, bark is considered to be the tissue of choice as source material for indexing, especially also as it provides cleaner nucleic acid preparations.[80] In the case of avocados, flowers contained a higher concentration of ASBV than did the leaves of infected trees.[81] CCCV was isolated from coconut palm roots, nut stalks, coconut husk, and the influorescence while high levels were found in pollen; only the coconut meat and milk and the trunk contained low concentrations.[82]

2. Purification

Slab polyacrylamide gel electrophoresis (PAGE) is a very efficient method for the fractionation of nucleic acids in partially purified plant extracts. In a two-step procedure which we have used extensively for viroids,[8,69,74,75] nucleic acid extracts are first run on a nondenaturing 5% polyacrylamide slab gel to separate the viroid (circular and linear forms comigrate) from host high and low molecular weight nucleic acids and colored material; extract from 100 to 125 g of plant material can be loaded on a 0.65 × 15 × 15 cm gel. After location of the viroid band by brief staining, the band is excised and recovered from the gel by electroelution followed by ethanol precipitation. In the second step, the viroid is further purified and fractionated on a 5% polyacrylamide slab gel containing 7 *M* urea, under which conditions the circular RNA molecules are separated from the faster migrating linear forms. In this second step, the partially purified viroid from 1 to 2 kg of plant material can be electrophoresed on one slab gel. The circular and linear molecules are again located by brief staining and recovered by electroelution and ethanol precipitation. Yields of viroids have varied from about 20 μg/kg of tomato leaves for CEV up to about 1 mg/kg of avocado leaves for ASBV.[83] An example of the two gel steps in the purification of both monomeric and dimeric ASBV is given in Figure 4.

For the rapid purification of circular viroid, small volumes of plant extract can be electrophoresed directly on a 7 *M* urea, 4% polyacrylamide slab gel, and the viroid band eluted after location by staining. RNA isolated in this way has proved to be pure enough for many purposes, e.g., for direct RNA sequencing or for the preparation of cDNA for cloning.[84] Where gels of less than 1 mm thickness are used, the viroid band can be eluted with high recovery simply by soaking the intact gel band overnight at room temperature in 0.5 *M* ammonium acetate, 10 m*M* EDTA, pH 8, 0.1% SDS, and recovering the RNA by ethanol precipitation.

A novel approach to the rapid, large-scale purification of viroids using Cs_2SO_4 gradient centrifugation and high-performance liquid chromatography (HPLC) was described by Colpan et al.[79] Leaves, stems, or roots of PSTV-infected tomato plants were homogenized with phenol, the nucleic acids in the aqueous phase recovered by ethanol precipitation and DNA removed by treatment with DNase I. The crude extract was then made to a density of 1.36 with $CsSO_4$ and centrifuged at 50,000 RPM for 16 hr. PSTV and other low molecular RNA banded just above the precipitate which consisted predominantly of high molecular weight RNA. The viroid and other low molecular weight RNAs were recovered by CTAB precipitation after dilution of the Cs_2SO_4 solution. Fractionation of these RNAs was achieved by HPLC on a weak anion exchanger linked to silica gel of optimized pore size (prepared by the authors). The elution was carried out in a 5 hr run by a salt gradient with the complete exclusion of divalent metal ions using a solution which contained EDTA. Because of the EDTA, a specially modified apparatus was necessary for HPLC in order to avoid corrosion of the stainless steel components. Electrophoretically pure PSTV was obtained in yields of about 0.5 to 1 mg/kg of plant material. The use of this procedure would be restricted to those laboratories which have the suitably modified equipment for HPLC. Unlike gel electrophoresis in the presence of 7 *M* urea, HPLC does not separate the circular and linear forms of the viroid.

B. Purification of Virusoids

Although virusoids can be readily isolated from nucleic acid extracts of plants as described above, they are traditionally isolated from purified virions where they constitute up to about 80% of the total viral RNA.[42,44,45,48] Fractionation of this RNA on a single polyacrylamide gel in the presence of 7 *M* urea is then sufficient for the isolation of the circular and linear forms.

C. Linear and Circular Forms in Purified Preparations of Viroids and Virusoids

All preparations of viroids and virusoids contain linear as well as circular molecules with the linear molecules varying from almost undetectable levels up to about 10% of the circular forms. We have found this range for four virusoids and three viroids in numerous preparations over several years. It has proven difficult to determine whether these linear molecules exist as such in vivo or arise, at least in part, by nicking of circular molecules during the extraction and purification procedures. The other unresolved problem is whether any linear molecules which exist in intact cells are the result of adventitious nicking by intracellular RNases or are the product of a step in the replication cycle. Some reports showed that the linear PSTV[85] and CSV[74] were infectious; presumably the linear RNAs were first ligated to the circular form before being replicated. Further discussion can be found in Sänger.[30]

D. Future Methods for Purification and Preparation of Viroids and Virusoids

1. Hybrid Selection

Although the monomeric and dimeric plus forms of viroids and virusoids can be readily detected and purified,[61,69] the isolation and characterization of minus forms and other greater-than-unit-length plus forms is more difficult due to the low levels and uncertainty of size.[61] A potential method for purifying such forms could be achieved by selective hybridization of plant extracts with single-stranded nucleic acid vectors, containing specific cloned, complementary viroid, or virusoid inserts, bound to a solid support.[86]

2. In Vitro Synthesis

The ability to prepare cDNA clones of viroids and virusoids in vectors containing highly efficient promoters for the RNA polymerases of bacteriophages SP6[87] and T7[88] makes the in vitro synthesis of specific RNA transcripts eminently feasible. In the case of ASBV, full-length monomeric plus and minus RNA transcripts could be easily produced in the tens of microgram range using dimeric plus and minus clones of ASBV in the vectors pSP6-4 and pSP6-5. Under the conditions used, there was rapid, specific self-processing of the dimeric RNA transcripts to the plus and minus monomers.[89] This method of artificially producing viroid and virusoid sequences may prove to be the only feasible means of obtaining sufficient amounts of minus RNA forms.

III. PHYSICIAL PROPERTIES

The ablity to obtain significant amounts of purified viroids and virusoids has allowed the physical and physicochemical investigations of these RNAs and the correlation of the results with sequence data. However, as considered below, there are still intriguing aspects of viroid structure for which we have little appreciation.

A. Circular Structure of Viroids and Virusoids is Revealed by Electron Microscopy

PSTV appears under native conditions in electron micrographs as rod-like molecules of an average length of 37 ± 6 nm.[29,90] From the molecular weight of PSTV and its sedimentation coefficient determined independently in solution, an axial ratio of 20:1 was calculated, a value in good agreement with the electron microscopic data. When PSTV was spread under denaturing conditions for electron microscopy, it was visible as covalently closed, single-stranded circles with contour lengths of 100 ± 6 nm,[29,90,91] a size consistent with its length of 359 nucleotides as determined by sequencing.[15]

Similar, but less extensive analyses by electron microscopy of CSV,[74] CEV and CPFV,[90] CCCV,[92] HSV,[93] ASBV,[3] and the virusoids of VTMoV,[42] SNMV,[44] LTSV,[45] and SCMoV[48] have shown that they are similar to PSTV in being single-stranded, covalently closed RNA molecules. The lengths of these molecules, when spread under denaturing conditions, were consistent with the exact sizes subsequently determined by sequencing.

B. Primary and Secondary Structures of Viroids and Virusoids

The sequencing of nine viroids and five virusoids (Tables 1 and 2, Figures 1 and 2) confirmed that they were all covalently closed single-stranded, circular RNA molecules. As yet, no modified bases have been detected.[15] The methods used in sequence determination are described below. In all cases it has been possible to predict a secondary structure for each RNA involving intramolecular base-paired regions separated by small internal loops to give a linear, rod-like structure (Figures 1 and 2). In practice, the quickest way to arrive at the possible secondary structure is to maximize the number of base pairs and to compare the structure with that of other related viroids and virusoids. The sequence of PSTV was determined in 1978 and its secondary structure predicted, aided by evidence indicating the sites of single-stranded regions which were susceptible to single-strand specific RNases and to bisulfite modification.[15] It was then straightforward to rapidly derive a secondary structure for related viroids on the basis of sequence homology and the maximization of the number of base pairs. However, in the case of the unrelated ASBV and the four virusoids, it was necessary to use the base pairing matrix procedure of Tinoco et al.[94] as a first step in the development of a secondary structure.

Calculation of the thermodynamically most stable structure can be used to add refinements to the predicted structure. The lowest free energy (ΔG) has been calculated for each structure using a set of stability constants, reaction enthalpies, and entropies for base pair formation.[29,95,96] The values used by different authors have varied but they depend on the type of base pair formed and the type of the neighboring base pair. In addition to the usual A:U and G:C base pairs, the G:U base pair is taken into account since its stability is similar to that of A:U base pairs.[29,96] Furthermore, in the case of the first base pairs on each side of a loop, the parameters depend on the size and kind of the loop. Computer programs are being developed which make use of a given set of parameters to calculate the most thermodynamically stable structure.[96-99]

C. Thermodynamic Investigations Provide Information on Viroid and Virusoid Structure

1. Structural Transitions of PSTV

The helix to random coil transition of nucleic acids can be brought about by an increase in temperature and is associated with a 30 to 40% increase in UV absorption. These transition

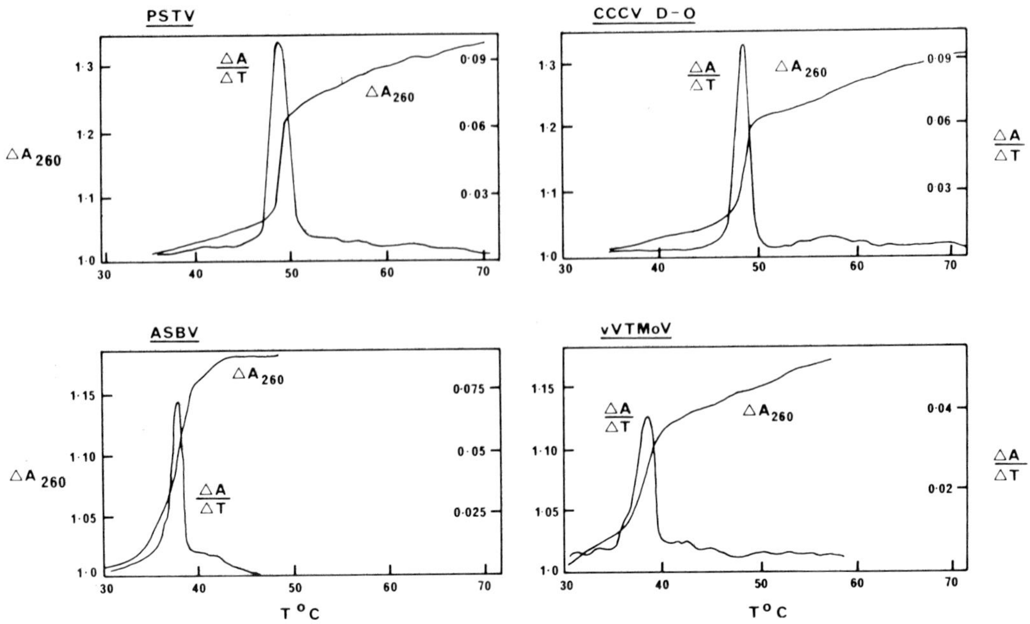

FIGURE 5. Denaturation curves at A_{260} nm and differentiated curves (ΔA/ΔT) in 10 m*M* NaCl, 1 m*M* sodium cacodylate, 0.1 m*M* EDTA, pH 6.8, of PSTV,[101] CCCV D-0,[102] ASBV,[96] and vVTMoV.[102]

or melting curves are best recorded with computer controlled spectrophotometers with microcuvettes which only require a few micrograms of nucleic acid.[100] The melting curves of PSTV,[101] CCCV,[38,95,102] ASBV,[96] and vVTMoV[102] in both their original and differentiated forms are given in Figure 5. Melting curves in the differentiated form are most useful because the occurrence of more than one transition can be recognized more easily and the evaluation of the midpoint temperature, T_m, and the half width, $\Delta T_{1/2}$, is easier.[103]

In the case of PSTV, the melting curve in the relatively low ionic strength buffer (1 m*M* Na cacodylate, 20 m*M* NaCl, 0.1 m*M* EDTA, pH 6.8) shows that most of the PSTV secondary structure dissociates in a highly cooperative process at intermediate temperatures over a short temperature range (Figure 5). After the main transition in which most of the structure has dissociated, particularly stable portions of the PSTV structure are still intact, but these dissociate at higher temperatures in transitions of much smaller amplitude and with broader width. Because of the low signal, these transitions were not resolvable by the equilibrium method just described. Resort was, therefore, made to the temperature jump technique whereby absorption changes can be monitored with temperature changes occurring over microseconds to minutes.[101] The method can resolve two or more transitions which are superimposed, but which have clearly different time courses.

On the basis of such thermodynamic data and before the complete sequence of PSTV was known, some essential features of its secondary structure were derived.[104] Models for the secondary structure were varied systematically and their melting curves were calculated applying known methods of statistical thermodynamics and the simulated melting curves were compared with those obtained experimentally. Best agreement with the melting curve obtained experimentally was for a secondary structure in which short double helixes and small internal loops were arranged in an unbranched series. This model agrees well with viroid structure as we now know it from all the sequence data.

From a knowledge of the sequence of PSTV, Riesner et al.[29] developed a model to explain their thermal denaturation data (Figure 6). In this model, the main transition where 80 to 90% of the hyperchromicity occurs is represented by steps S.1 and S.2, in which the left half of the molecule denatures first (S.1) as the temperature is raised to give a metastable

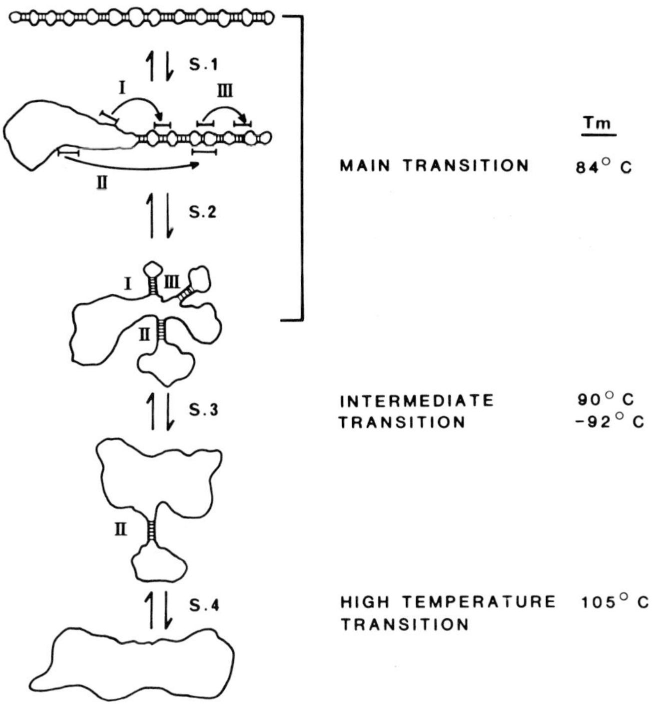

FIGURE 6. Mechanism of the denaturation of PSTV with increasing temperature.[29,101] See Table 4 and text for details of the formation of the hairpin loops I, II, and III during the main transition. The T_m values of the three transitions are at 1 *M* ionic strength.

form in which the right half results in the formation of three new stable hairpins, I, II, and III (S.2), between complementary segments in distant parts of the molecule. The formation of these three hairpins presumably enhances the cooperativity (sharpness) of the main transition.

As the temperature increases further, these hairpins denature in separate thermal transitions (S.3 and S.4) with the order of melting being determined by their relative stabilities. It is the melting of these hairpins, which are not part of the native structure of PSTV, that provides the hyperchromicity above the main transition. The three hairpins were originally detected on the basis of fast kinetic measurements using the temperature jump technique[101] as three extra transitions. The thermodynamic data obtained in this work and the sequence of PSTV allowed the three hairpins to be clearly defined. The intermediate structures formed after S.2 and S.3 are stable enough to be observed by electron microscopy.[29,38]

In summary, therefore, PSTV combines the sharpness of the main thermal transition, i.e., the cooperativity of double-stranded nucleic acids, with the low T_m of single-stranded, clover-leaf like tRNA structures.[38] The whole linear arrangement of helixes and loops of the viroid in combination with circularity serves for the maintenance of cooperativity, and the many internal loops decrease the melting temperature.

2. Structural Transitions of Other Viroids and of Virusoids

Thermodynamic investigations of viroids other than PSTV and of the virusoids have shown varying degrees of similarity with the structure and structural transitions of PSTV. For example, the melting profile of CCCV (Figure 5) shows that the T_m of the main transition, cooperativity of melting, and the hyperchromicity were very similar for PSTV, CEV, and CSV (Figure 5, Table 3). Although the T_m values for CCCV, PSTV, CEV, and CSV were

Table 3
PROPERTIES OF PROPOSED SECONDARY STRUCTURES OF VIROIDS AND VIRUSOIDS

RNA	No. of nucleotides	Nucleotides base paired (%)	G:C base pairs as % of total	ΔG (kJ/mol at 25°C)	ΔG / No. of nucleotides	Thermodynamics[a] T_m (°C)	Thermodynamics[a] $\Delta T_{1/2}$ (°C)	Ref.
Viroids								
ASBV	247	67	34	−280	−1.13	37.5	1.5	4,32,96
CCCV	246	66	68	−320	−1.30	49	1.2	11,95
CEV	371	67	58	−590	−1.59	51	1.0	8,95
CSV	356	70	52	−540	−1.52	48	1.1	5,95
PSTV	359	70	58	−600	−1.67	51	0.9	15,95
Virusoids								
vLTSV	324	72	56	−440	−1.36	—	—	39
vSNMV 1[b]	378	73	55	−445	−1.21	—	—	40
2[c]	378	66	57	−540	−1.43	38	2.8	32,95
vVTMoV 1[b]	366/367	67	58	−350	−0.96	—	—	40
2[c]	366/367	64	56	−480	−1.32	38	2.0	32,95
Random sequences[d]	359	59	58	−456	−1.27	36 ± 5	>5	32,95

[a] Data obtained at ionic strength 0.011 Na^+, pH 6.8.
[b] For completely rod-like molecule as in Figure 2.
[c] For rod-like molecule with single bifurcation in proposed structure.
[d] Average values for five different sequences with same number of nucleotides and base composition as for PSTV.

all within a range of 3°C at the same ionic strength (Table 3), the denaturation of CCCV was highly cooperative only in low ionic strength, losing its cooperativity with increasing ionic strength,[102] a behavior not found with the other viroids.[104]

In the PSTV-like viroids, there is variation in the ability to form the three stable hairpins shown by PSTV on denaturation (Figure 6). All the viroids can form the stable hairpin I because of the highly conserved central sequence (Figures 1 and 7, Table 4). However, CCCV and HSV cannot form hairpins II and III because the appropriate complementary sequences are not present, while CSV, CEV, TASV, and TPMV can form hairpins I and II but not III (Table 4). Hence, the hyperchromicity that occurs after the main transition in the melting curves of PSTV, CCCV (Figure 5), and other viroids must have a contribution from the denaturation of other, undefined structures.

The more limited thermodynamic analyses of ASBV[32,96] have shown a similar cooperative melting profile except for a lower T_m and a higher $\Delta T_{1/2}$, as compared to other viroids (Table 3, Figure 5). This may, in part, be explained by the A:U-rich nature of ASBV compared to all other viroids and virusoids which are G:C rich (Table 3). This would be expected to have a pronounced effect on the secondary structure as the difference in stability of A:U and G:C base pairs in RNA greatly exceeds the difference in stability of dA:dT and dC:dG base pairs in DNA.[106]

Several thermodynamic properties indicate structural differences between viroids and virusoids (Table 3). The T_m of the main transition for the VTMoV and SNMV virusoids was 10°C lower than found with viroids and the cooperativity was markedly lower as indicated by the relative widths ($\Delta T_{1/2}$) of the differentiated main transition of the melting curves (Figure 5, Table 3).Further, on the basis of the analysis of the sedimentation coefficients of CCCV, PSTV, the virusoids of VTMoV and SNMV, and the double-stranded form of cucumber mosaic virus satellite RNA (a linear molecule of 335 base pairs[107]), Riesner et al.[108] showed that the virusoids were more flexible than viroids which themselves were more

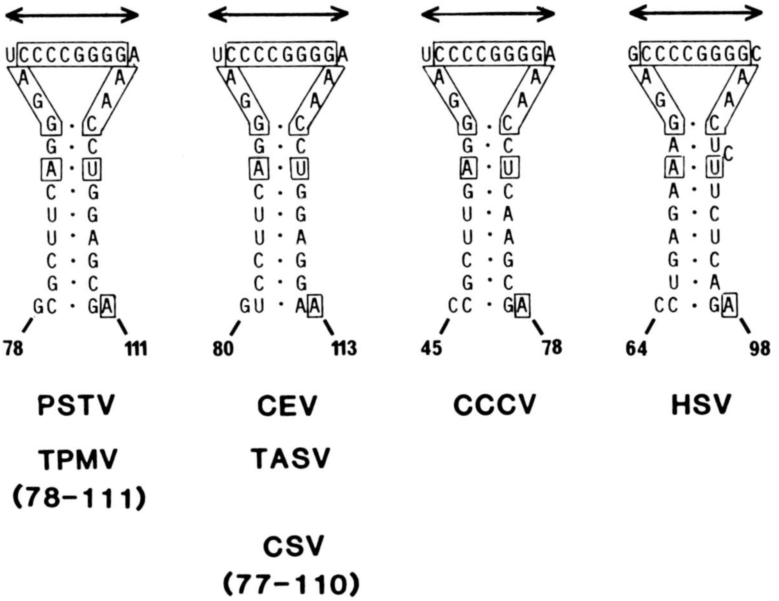

FIGURE 7. Alternative secondary structures in the central conserved region for the PSTV-like viroids[105] which correspond to hairpin loop I.[101] The 9 bp stem (Figure 6) is terminated by a loop containing a 10 nucleotide self-complementary sequence (↔). The CCCGG sequence in the proposed loop corresponds to one strand of the strictly conserved bulged helix (Figure 13) which would thus be disrupted by the formation of base pair stemloop structures. The boxed nucleotides are conserved in all PSTV-like viroids. (From Keese, P. and Symons, R. H., *Proc. Natl. Acad. Sci. U.S.A.*, 82, 4582,1985. With permission.)

Table 4
POTENTIAL ABILITY TO FORM HAIRPIN LOOPS DURING THERMAL DENATURATION OF VIROIDS

		No. of nucleotides in stem or loop[a]						
Loop no.		**PSTV**	**CSV**	**CEV**	**CCCV**	**HSV**	**TASV**	**TPMV**
I	Stem[b]	18	18	18	18	18	18	18
	Loop	14	14	14	14	14	14	14
II	Stem[b]	20	22	26	—[c]	—	28	22
	Loop	82	76	75	—	—	75	88
III	Stem[b]	18	—	—	—	—	—	—
	Loop	26	—	—	—	—	—	—

[a] Derived from sequences in Figure 1.
[b] Stem length greater than six base pairs.
[c] Indicates that stem or loop cannot form.

flexible than the dsRNA. Calculations indicated that the largest viroid studied (a CCCV dimer of 600 nucleotides) could bend in solution to more than 25% of a circle. However, this distinction between viroids and virusoids is not entirely clear cut since vLTSV is more stable than either vVTMoV or vSNMV as indicated by a higher T_m[45] and lower $\Delta G/N$ (Table 3[39]), while ASBV, and to a lesser extent, CCCV, appear less stable than the other viroids.[32]

A summary of some of the other properties of the proposed secondary structures of five viroids and three virusoids is given in Table 3. The percentage of nucleotides base paired varies in the narrow range from 66 to 73%. Circular RNA molecules of random sequence and the same size as viroids were calculated to contain about 55% of their nucleotides base paired.[29] In addition, all RNAs, except ASBV, contain a similar proportion of G:C base pairs. The thermodynamic stabilities (△G) of the proposed structures range from −280 kJ/mol for ASBV to −610 kJ/mol for PSTV. These values can be more meaningfully compared by dividing the △G value by the number of nucleotides (△G/N, Table 3). Although CEV, CSV, and PSTV are clearly more stable on the basis of these data than the virusoids, the situation is different with ASBV and CCCV where the △G/N values are in the range found for virusoids. This contrasts with the thermal denaturation profile of CCCV (Figure 5, Table 3) which is comparable with that of the viroids. Hence, the △G calculations based on various stability parameters for RNA, such as base pairing, hairpin loop formation, internal loops, bulge loops, and base pairing within a circular molecule, do not provide a complete overall assessment of the relative stability of these molecules.

The question arises as to whether there is, in viroids and virusoids, tertiary structure similar to that found in other single-stranded RNAs such as tRNA, 5S RNA, and rRNAs. The L-shaped tRNA is the best studied example of a nucleic acid with pronounced tertiary structure.[109] Different experiments have indicated that the extended form of viroids is actually present in solution. These experiments have included electron microscopic and hydrodynamic studies, studies with dye binding to helical regions, oligonucleotide and tRNA binding to the unpaired regions, enzymatic digestion in the loops, and chemical modification of unpaired cytosines.[15,29,38] In spite of this evidence, the possibility must be considered that there are additional intramolecular interactions which have yet to be discovered.

An interesting example is the specific cross-linking induced by irradiation with UV light in PSTV and HeLa 5S rRNA.[110] Intramolecular cross-linking in each of these molecules occurred at a single major site in an identical eight base context, with the sequence GGGAA on one side and the sequence UAC on the other. Both cross-links connect bases which are not thought to be involved in conventional hydrogen bonding, in the center of regions usually depicted as single-stranded loops flanked by short helical regions. All sequencing data indicate that in PSTV, nucleotide G98 (middle nucleotide of GGGAA) cross-links with U260 in the highly conserved central region of the viroid sequence, while in HeLa 5S RNA, G101 cross-links with U78, bases which are invariant among eukaryotic 5S rRNA molecules.[111] The results clearly indicate a single, activated tertiary structure within conserved regions of both molecules. Such a structure has not been predicted from previous thermodynamic experiments which were probably too insensitive to detect this localized tertiary structure and possibly others which exist in native molecules.

Clearly, a thorough understanding of all elements of secondary and tertiary structure in viroids and virusoids will be crucial to relating structure to function in these molecules. This includes consideration of factors such as nonstandard base pairs. For example, G:A base pairs have been found in both RNA and DNA[109] and have been postulated for viroids.[112] Also, C:A base pairs have been proposed for tRNA and may, in the right conformation, be more stable than A:U and G:U base pairs.[113]

IV. SEQUENCE AND SECONDARY STRUCTURE

A. Sequence Determination of Viroids and Virusoids

The usual starting material for sequence analysis is the purified, infectious circular viroid or virusoid. Although the linear monomeric forms of PSTV and CSV are infectious when isolated from infected plants,[30,74,85] the ends are heterogeneous, possibly because of nonspecific cleavage of the circular molecules. This makes sequencing difficult by the direct

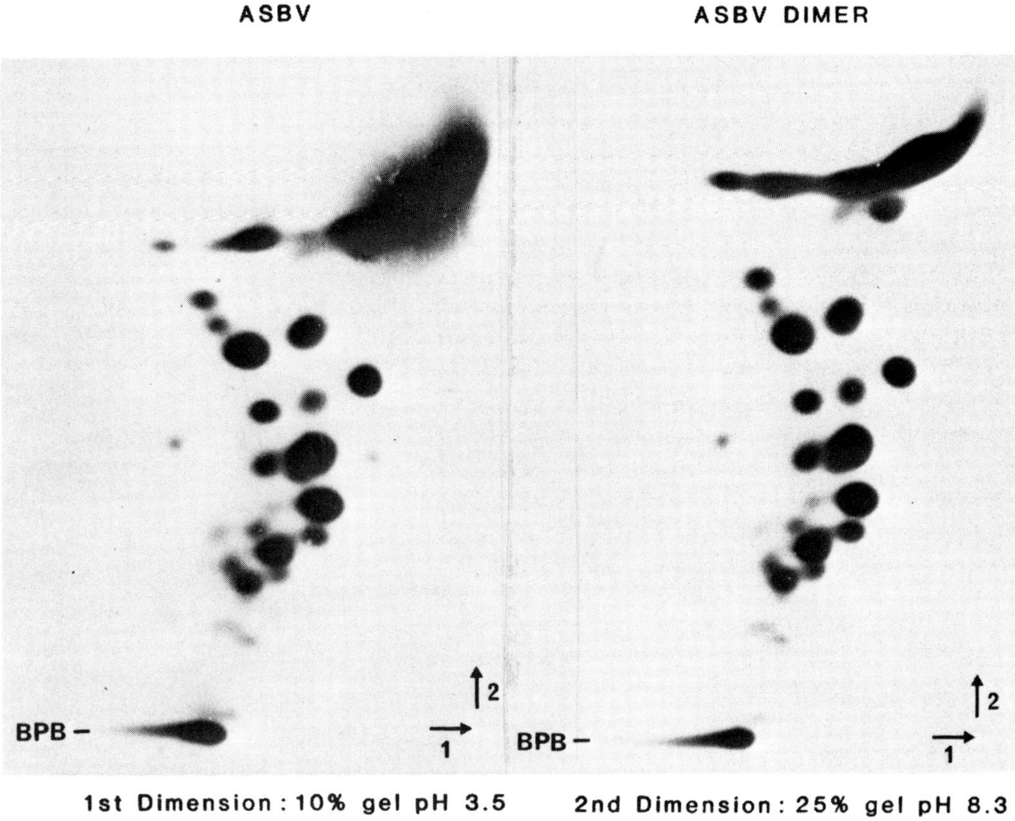

FIGURE 8. Two-dimensional fingerprinting of pancreatic RNase digests of ASBV (A) and dimer ASBV (B). The detailed procedure is described in Reference 68. The first dimension was by electrophoresis in a 10% polyacrylamide gel at pH 3.5 and the second dimension was in a 25% polyacrylamide gel at pH 8.3. BPB is the position of the bromophenol blue marker in the second dimension. The large radioactive band at the top of each gel is mostly unincorporated γ-^{32}P-ATP.

method which requires a unique 5′- and 3′-end. There are, at present, four approaches employed to compare viroids and virusoids and to determine their complete sequences.

1. Fingerprint Analysis

Two-dimensional fingerprinting is a rapid and sensitive method for comparing two or more RNAs and was used extensively in the initial characterization of viroids.[15,68,114-118] Purified RNA is digested to completion with a base specific RNase such as RNase T_1, specific for guanine, RNase U_2, specific for adenine at pH 3.5, or pancreatic RNase A, specific for pyrimidines. Since all these RNases cleave to leave a 5′-hydroxyl terminal nucleotide, the mixture of RNA fragments can then be 5′-labeled with ^{32}P using γ-^{32}P-ATP and phage T4 polynucleotide kinase. Alternatively, viroids can be labeled on C nucleotides with ^{125}I prior to enzymic digestion. These labeled oligonucleotides are then separated on the basis of size and charge differences by two-dimensional paper chromatography[15] or by two-dimensional PAGE[119,120] to give a characteristic fingerprint.[15,68,114-118] The method is very sensitive for the detection of one or only a few base changes in a viroid or virusoid since even a single base change can lead to the disappearance of one oligonucleotide spot and the appearance of another.

Each oligonucleotide after purification can be sequenced as originally performed for PSTV,[15] but this method is very tedious for the accumulation of sequence data. However, fingerprinting can still play a role in rapid diagnosis as, for example, in the confirmation of the identity of the dimeric forms of ASBV[69] (Figure 8) and of CCCV.[68] Fingerprint

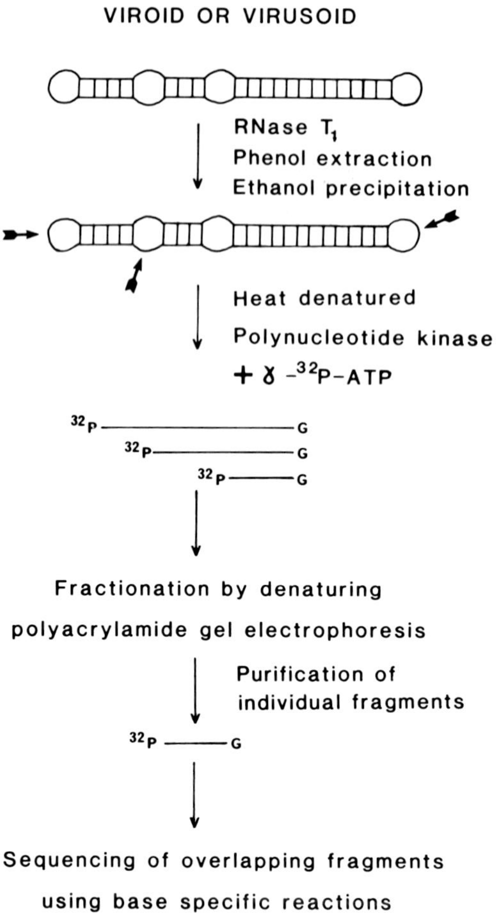

FIGURE 9. Strategy for obtaining overlapping linear fragments with specific termini for sequencing from circular, single-stranded viroid or virusoid.[5,8,11,39,40] A set of overlapping fragments is produced by partial enzymatic digestion of the RNA. This is usually performed under conditions of high salt concentration and low temperature to limit the number of potential cleavage sites (e.g., as shown by arrows). After heat denaturation of the nicked RNA, the 5′-ends are labeled with γ-^{32}P-ATP and polynucleotide kinase and the fragments fractionated by PAGE on a denaturing gel. Individual fragments are located by autoradiography, eluted, and sequenced using base-specific enzymic reactions.[5,112,121-127]

analysis provided the first conclusive test to distinguish CEV and PSTV which share similar sizes and biological properties.[114]

2. *Direct RNA Sequencing*

This technique (Figure 9) is used after prior partial cleavage of the RNA with base specific RNases followed by 5′-end labeling with γ-^{32}P-ATP and T4 polynucleotide kinase or 3′-end labeling with T4 RNA ligase and ^{32}P-pCp.[121] Ideally, the partial cleavage should produce a series of overlapping fragments which are representative of the whole molecule (Figure 9). After end labeling, the fragments are separated by PAGE, located by autoradiography, eluted from the gel, and individual fragments sequenced directly with base-specific enzymes[5,122] or by specific chemical cleavage in the case of the 3′-labeled fragments.[123]

There are two difficulties with the use of base-specific enzymes for RNA sequencing. First, although purines can be readily distinguished using RNase T_1 and RNase U_2 at pH 3.5,[122,124] there are no enzymes completely satisfactory for deciphering pyrimidines. Enzymes that have been commonly used include RNase PhyI,[125] RNase PhyM,[126] extracellular RNase from *Bacillus cereus*,[127] and chicken liver RNase.[128] However, none of these except possibly RNase PhyM provide uniform enough digestions to allow unambiguous distinction between C and U nucleotides. Second, internal secondary structure of fragments from G:C-rich RNAs such as viroids and virusoids can be stable enough so that fragments are difficult to resolve on sequencing gels due to band compression[129] even in the presence of 7 *M* urea. The use of a stronger denaturant such as 98% formamide[5] or 40% formamide and 7 *M* urea has aided in overcoming this problem.

The chemical method of sequencing[123] can satisfactorily distinguish all four bases. In addition, after the various chemical treatments, the fragments sequenced by this method no longer form tight secondary structures. However, chemical cleavage of RNA is more laborious than enzymatic sequencing and it can only be used for sequencing fragments that are 3′-labeled. Chemical sequencing of 5′-labeled fragments does not produce readable sequencing gels.

In practice, direct enzymatic sequencing is rapid and can provide about 80% or more of the sequence of a viroid or virusoid in a few days starting with only a few micrograms of RNA. However, finalizing the sequence involves considerably more effort with repeated reactions under a variety of conditions to resolve ambiguities between C and U nucleotides and band compressions. In addition, any region of the RNA molecule which is very susceptible to RNase digestion can be poorly represented in the partial RNase digests.[4] The complete sequence of the circular viroid or virusoid is finally obtained from overlaps of the sequences of a number of fragments.

3. *Primer-Directed Sequence Analysis*

An approach with considerable potential for the rapid examination of a number of isolates of the same viroid is to use an oligonucleotide primer for the synthesis of cDNA with reverse transcriptase. Sequencing of the DNA can be done by the chemical method of Maxam and Gilbert[130] starting with a 5′-^{32}P-labeled primer or by the dideoxynucleotide chain termination technique.[131-132] This approach has been used for the rapid examination of a number of different PSTV isolates with three chemically synthesized oligonucleotide primers which hybridize to different parts of the PSTV molecule.[16] In the case of CEV, the use of synthetic primers for dideoxy sequencing has provided useful initial data, but extensive band compression has prevented its use for accurate sequence determination.[10,84,112] In the selection of DNA primers for this approach, it is important to make use of regions of the viroid or virusoid molecule containing a minimum of secondary structure and to have primers of sufficient length (preferably 15 nucleotides or more) to ensure a stable hybrid for initiation of cDNA synthesis. DNA primers can also be isolated by restriction enzyme digestion of recombinant cDNA clones of viroids and virusoids and have been used for the confirmation of part of the sequence of ASBV,[4] CEV,[8] vVTMoV, and vSNMV.[40]

4. *Sequencing of Viroids and Virusoids Using Cloned cDNA*

The recognition that naturally occurring isolates of CEV can be complex mixtures of a number of sequence variants[9,10] has very important implications for sequence analysis. Direct RNA sequencing and primer-directed sequencing cannot resolve mixtures of two or more sequence variants and will not detect minor amounts of sequence variants in the presence of a single dominant variant.[8,9] The most feasible approach to overcome these difficulties is to sequence a number of full-length cDNA clones prepared from the purified isolate of viroid or virusoid. The complexity of the isolate will determine the number of clones that have to be sequenced to give a measure of the number of sequence variants and their relative distribution.

Both partial- and full-length cDNA clones of viroids and virusoids have been reported.[9,10,14,66,77,133-136] Cloning into the ssDNA of bacteriophage M13 vectors allows rapid

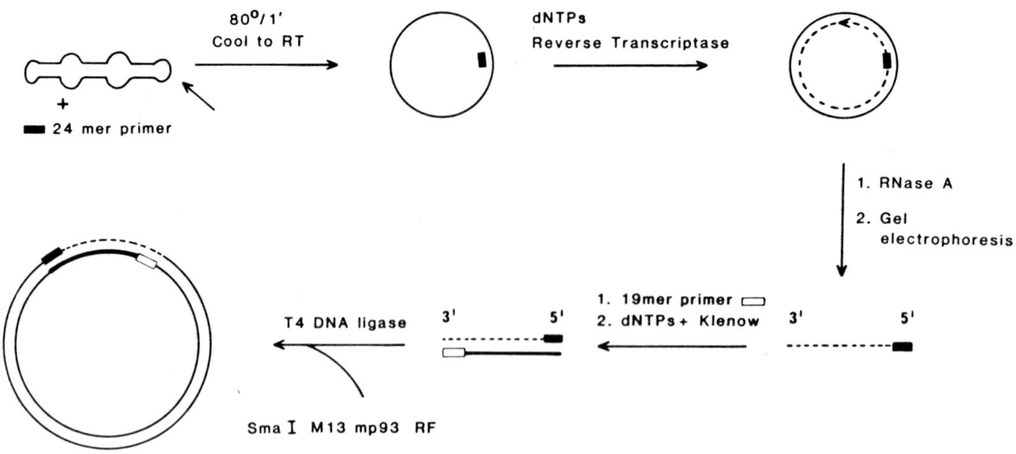

FIGURE 10. Outline of the procedure used in the construction of full-length cDNA clones of CEV.[10] The site of priming of the 24 nucleotide DNA primer at the right-hand end of the predicted native structure of CEV is indicated by an arrow. (From Visvader, J. E. and Symons, R. H., *Nucleic Acids Res.*, 13, 2907, 1985. With permission.)

sequence determination by the dideoxy chain termination procedure.[137] The preparation of full-length clones of viroids[10,14] has the distinct advantage of being able to identify individual RNA variants and has been successfully used to resolve the highly complex mixtures of two CEV isolates.[10]

In the preparation of full-length cDNA clones of HSV, Ohno et al.[14] started with linear molecules and used the procedure of Okayama and Berg,[138] which does not require prior knowledge of the viroid sequence. We have found an alternative approach, which requires at least partial knowledge of the viroid sequence, to be very successful for the rapid cloning of CEV using only 0.5 to 2.0 μg of circular viroid.[10] The procedure for the cloning of CEV in the phage M13mp93 vector is outlined in Figure 10. It requires the sequential use of two synthetic oligonucleotide primers to prepare full-length cDNA which is then blunt-end ligated into the replicative form of the vector. Infection of bacteria produces clones of progeny M13 phage containing either plus (viroid) or minus (complementary) sequences, each of which can usually be sequenced fully in a single reaction by the dideoxy nucleotide chain termination technique.[137]

B. Structure of Viroids

1. Structural Domains of Viroids

The viroids examined to date range in size from 246 to 375 nucleotides (Table 1); CV and CVaV have still to be characterized. On the basis of sequence and structural homology, the sequenced viroids can be divided into two classes. One class consists only of ASBV, while the remaining viroids constitute a second class since they are all related to the best characterized viroid, PSTV. Although ASBV possesses a predicted rod-like structure (Figure 1) with properties similar to other viroids, it has a higher A:U content, a higher ΔG/size ratio (Table 3), and lower sequence homology (20%) with any other viroid. The PSTV-like viroids (Table 3) share at least 35% sequence homology between any two members and they all contain a conserved, central, U-bulged helix, $\begin{matrix}\mathrm{CC\ \ CCGG}\\ \mathrm{GGUGGCC}\end{matrix}$, and an oligo ($A_{5\text{-}6}$) segment that occurs in the same relative position about 35 nucleotides 5′ of the central conserved sequence, $\mathrm{GGA}_{\mathrm{G}}^{\mathrm{U}}\mathrm{CCCCGGGG}_{\mathrm{C}}^{\mathrm{A}}\mathrm{AAC}$.

From an examination of sequence homology within more than 40 PSTV-like viroid variants, a model of viroid structure applicable to all viroids, except ASBV, has been developed[105]

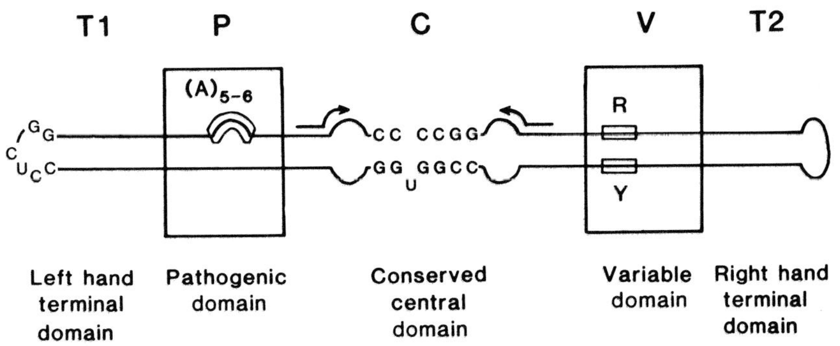

FIGURE 11. Model of five viroid domains (T1, P, C, V, T2) determined from sequence homologies between viroids.[105] The arrows depict an inverted repeat sequence which can form the stem loop I of Figure 7 and Table 4. R, Y: a short oligo-purine, oligo-pyrimidine helix. The dash in the end loop of the T1 domain represents a U nucleotide in CCCV and HSV but is absent in the other PSTV-like viroids.

in which five structural domains are distinquished (Figures 11 and 12). From this model it is possible to examine evolutionary relationships and draw functional inferences. The general features of these domains are summarized as follows.

C Domain: This conserved central region is centered around the strictly conserved bulged helix, $\begin{matrix}CC\ \ CCGG\\ GGUGGCC\end{matrix}$. may represent an important control region in viroid replication by signaling functional change through structural alterations.

P Domain: This region is associated with symptom expression and is characterized by an oligo ($A_{5\text{-}6}$) sequence present in all PSTV-like viroids.

V Domain: This region shows the greatest sequence varibility between closely related viroids.

T Domains: On the basis of sequence homologies, the terminal regions are considered to have undergone intermolecular RNA exchange between viroids to give rise to new, chimeric viroid species. Although the functional role of these domains is unclear, the evidence for these exchanges suggests a role for RNA rearrangements in the origin and evolution of viroids.

2. Control Function of the Highly Conserved C Domain

The central highly conserved domain of viroids consists of about 100 nucleotides rather than 44 or 56 nucleotides as originally described.[11,17] An example of the high degree of sequence conservation is shown by the 99% sequence homology between PSTV and TPMV in the C domain, although showing only 73% overall sequence homology (Table 5). The C domain of CCCV D-O, which constitutes 40% of the molecule, shows 70% sequence homology with PSTV, a value greater than the 65% sequence homology for this domain between PSTV and two other closely related viroids, CEV and TASV. These comparisons support the association of CCCV D-O with the PSTV-like group of viroids.

Although HSV shares a less closely related C domain (Table 5), it shows several sequence and structural homologies with the C domain of other viroids. These include a common, U-bulged helix, $\begin{matrix}CC\ \ CCGG\\ GGUGGCC\end{matrix}$, which is postulated to occur in the native viroid and the top strand of which forms part of a larger 16 nucleotide conserved sequence, $GGA^{U}_{G}CCCCGGGG^{A}_{C}AAC$. In addition, an alternative structure (Figure 7) is possible for this conserved bulged helix (Figure 13) which corresponds to the stem loop I, and this forms

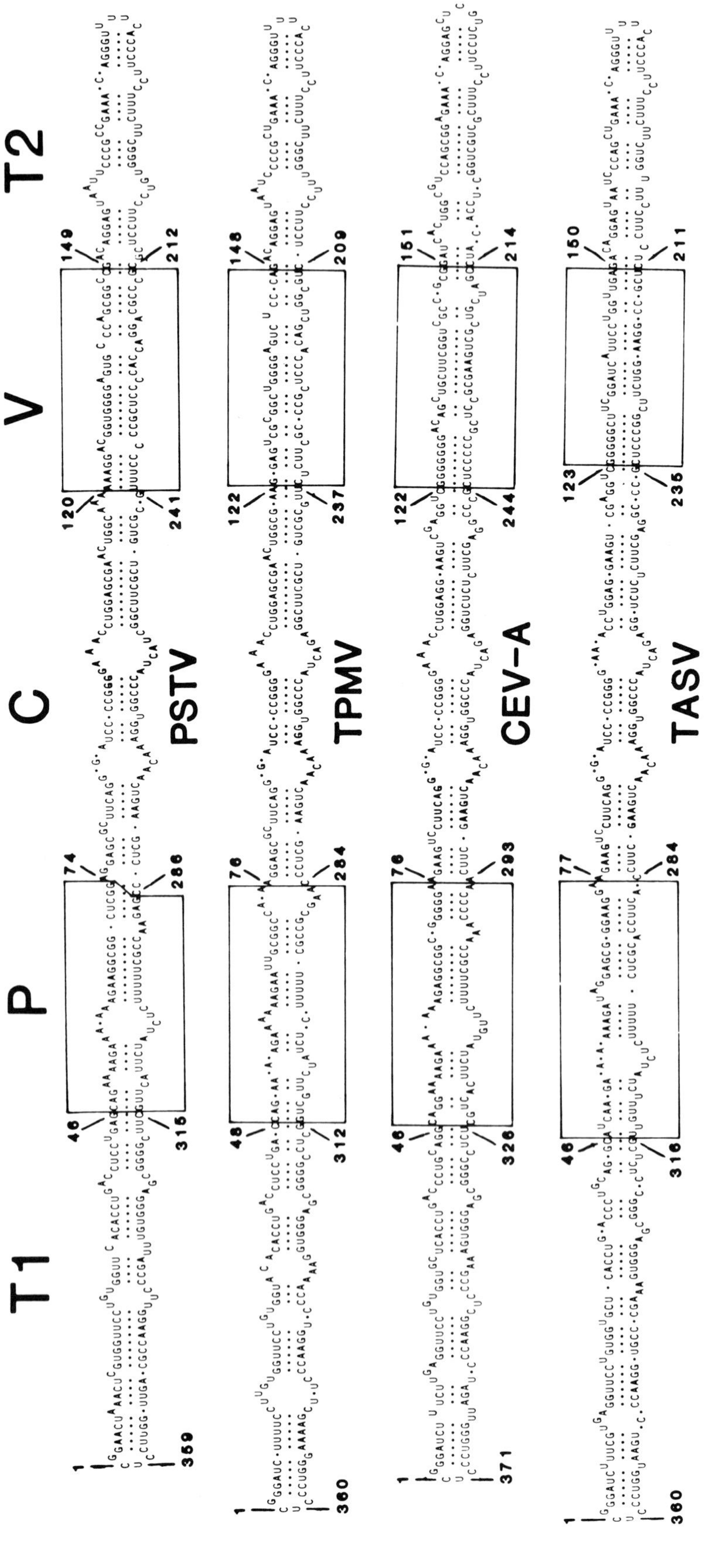
T1
P
C
V
T2
PSTV
TPMV
CEV–A
TASV

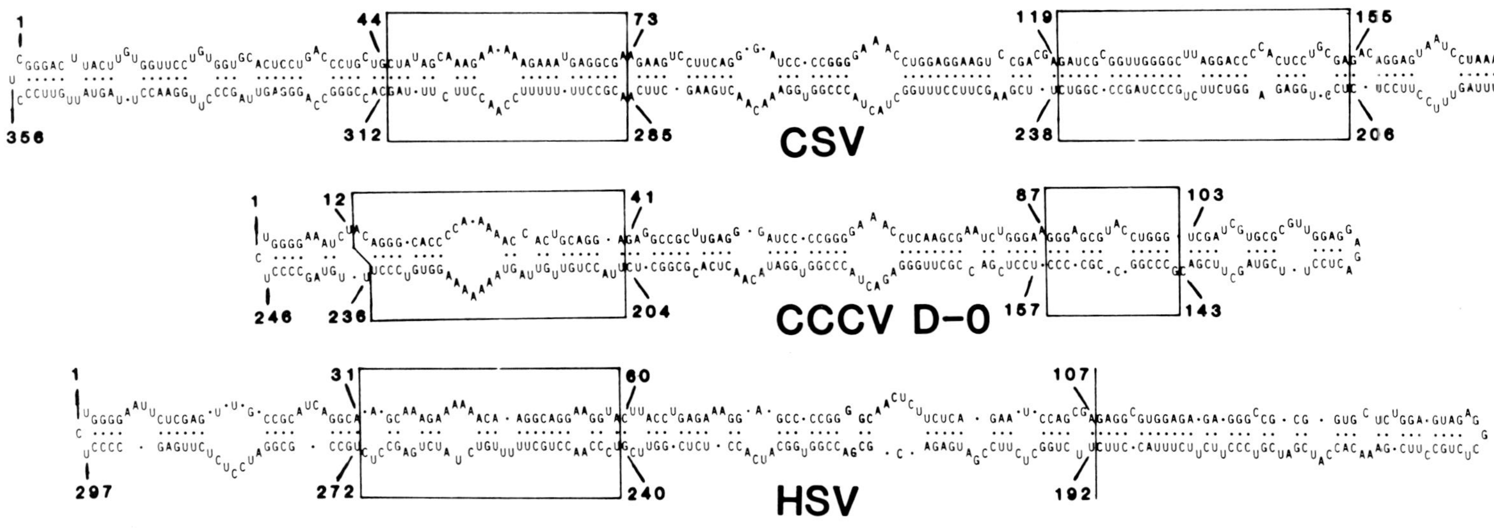

FIGURE 12. Domain boundaries defined for each viroid.[105] In pair-wise sequence comparisons of viroids containing highly homologous C domains (e.g., between PSTV, TPMV, and CCCV, or between CEV-A, TASV, and CSV), the P and V boundaries are defined by the significantly lower sequence homology which starts 5 to 9 nucleotides 5′ and 7 to 15 nucleotides 3′ of the inverted repeat sequence (Figure 11) in the C domain. Similarly, when comparing PSTV, TPMV, TASV, and CSV, a change from low homology in the V domain to high homology in the T2 domain defines the boundary for these two domains. However, the boundary of the V and T2 domains of HSV cannot be determined due to the present lack of homologous viroid sequences. Therefore, only the boundary between C and V domains of HSV is given. The P domain, with a region containing the conserved oligo-A sequence, is flanked by regions with greater variability and has its borders based on homologies between the P region of HSV and other viroids such as PSTV and by certain pair-wise comparisons such as CEV-A and TASV where there is significant change from relatively low sequence homology in the P domain to higher homology in the adjacent T1 and C domains.

during the thermal denaturation of PSTV, CEV, CSV, and CCCV[29,38] (see Section III.C).By such a scheme, the highly conserved CCCCGGGG sequence would form part of a self-complementary loop of a nine base pair stem (Figure 7) which gives a structure mutually exclusive to the bulged helix. Despite many sequence differences, the nine base pair stem can be formed for all viroids including HSV which differs only by the presence of a single, nonbase-paired C nucleotide (Figure 7).

It has been proposed[105] that both mutually exclusive structures are important in viroid function and that the structural switching from the native form to the nine base pair stem structure controls a switching in function for this region. For example, one function postulated for the C domain is a role in processing of viroid RNA replicative intermediates which is based on infectivity studies. cDNA clones and in vitro-synthesized RNA transcripts from cDNA clones of PSTV, CEV, and HSV have been shown to be infectious.[112,135,139] Infectivity of less-than-dimeric-length viroid clones[112,136,139] had been correlated with a partial duplication of at least 11 nucleotides (underlined) of the central conserved 16 nucleotide sequence, $\underline{\mathtt{GGA}^{\mathtt{U}}_{\mathtt{G}}\mathtt{CCCCGGGG}}^{\mathtt{A}}_{\mathtt{C}}\mathtt{AAC}$. It has been postulated that in vivo processing occurs within the 11-nucleotide repeat to generate a unit-length monomeric viroid free of any vector sequences.[112] Mutagenesis studies indicate that the processing site is within the four adjacent G nucleotides of this repeat.[112]

An alternative function for this region which may require a structural signal different from that involved in processing is based on the similarity of the bulged helix (Figure 13) with several protein binding sites,[140] in particular the ribosomal protein L18 binding site of 5S RNAs. It has been proposed that the single unpaired nucleotide (U in plant 5S RNAs) and the adjacent G nucleotide are crucial to the interaction with L18.[140,141] This analogous region in viroids may be a site for RNA polymerase binding and/or the initiation of synthesis of the minus strand to initiate the replication process. Therefore, structural switching in the C domain could control switching between two phases of the replication cycle, namely, transcription and processing.

3. Association of Pathogenicity with the P Domain

Many sequence variants of PSTV and CEV have now been characterized. Six variants of PSTV have been shown to differ in both sequence and severity of symptoms when propagated on tomato.[16,117,118] The nucleotide differences are confined to the P and V domains (Table 6), but only sequence changes in the P domain are correlated with variation in symptom expression;[16,118] this region has been defined by Sänger[16] as a virulence modulating domain (VM).

Interestingly, 17 sequence variants of CEV which differ by up to 29 nucleotides between any two variants[108] have most of these sequence differences located in the P and V domains (Table 6) and hence in the same relative positions as the PSTV variants. The CEV variants form two classes of sequences which differ by a minimum of 23 nucleotides in a total of 370 to 375 nucleotides.[10] These two classes correlate with two biologically distinct groups when propagated on tomato plants where one produces severe and the other mild symptoms. Host factors must also play a role since these two sequence classes of CEV do not correlate with differences in symptom expression when propagated on citrus.[10] Since some isolates of CEV are complex mixtures of sequence variants (CEV-DE30, CEV-J[10]), individual variants need to be tested for infectivity and symptom expression through the use of full-length clones.

Whether or not both the P and V domains of CEV affect symptom expression remains to be determined. One approach to investigate this further is to construct chimeras of viroids with various combinations of P and V domains. In the first successful experiment, an infectious CEV chimera was constructed from two CEV clones such that the left half of a

Table 5
SEQUENCE HOMOLOGY BETWEEN DOMAINS OF DIFFERENT VIROIDS[105]

Viroids used for pair-wise comparison		Sequence homology (%)					
		Domains					
1	2	T1	P	C	V	T2	Overall
TASV	CEV-A	91	54	99	49	46	73
	PSTV	67	59	65	30	90	64
TPMV	PSTV	67	73	94	42	95	76
	CEV-A	80	70	69	29	37	60
CCCV D-O	PSTV	25	14	70	37	27	38
	HSV	52	33	42	ID	ID	39
HSV	PSTV	23	58	35	ID	ID	35
CSV	CEV-A	77	42	82	28	38	59
	PSTV	69	49	71	31	81	61
CEV-A	PSTV	62	71	65	31	38	55

Note: Sequence homology was determined by the best alignment, allowing for additions and deletions, but constrained by the requirement of a match consisting of a minimum of three consecutive nucleotides. % Sequence homology = number of matching nucleotides in both sequences/total number of nucleotides in both sequences × 100.

ID = insufficient data.

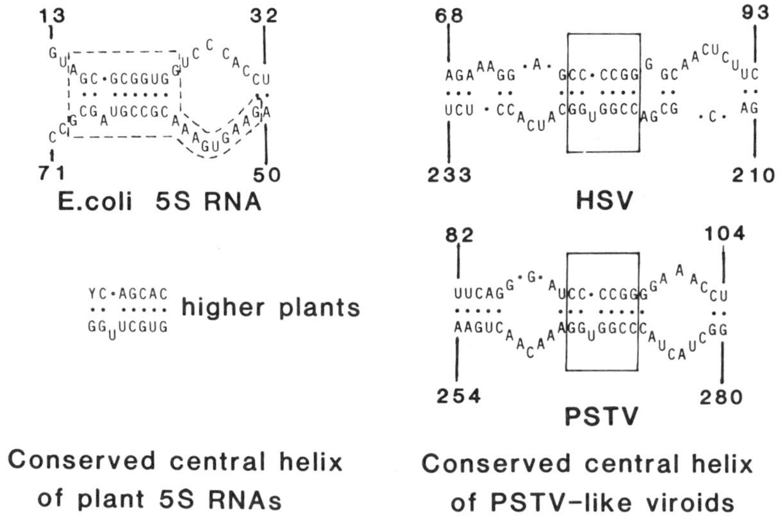

FIGURE 13. Structural homology between the ribosomal protein L18 binding site on 5S RNA of *Escherichia coli*[140] (dashed box) and the central conserved bulged helix of viroids (solid box in HSV and PSTV). The central conserved bulged helix of the 5S RNAs of higher plants[140] is given; Y represents a pyrimidine.

Table 6
LOCATION OF SEQUENCE DIFFERENCES BETWEEN 17 SEQUENCE VARIANTS OF CEV AND 6 SEQUENCE VARIANTS OF PSTV[a]

	No. of nucleotides which vary in viroid domain					
Viroid	**T1**	**P**	**C**	**V**	**T2**	**Ref.**
CEV	2 (92)	16 (61)	5 (97)	20 (57)	0 (64)	6,8-10
PSTV	0 (91)	12 (55)	0 (93)	2 (56)	0 (64)	15,16

[a] Number of nucleotides in each domain given in parentheses (see Figure 12).

variant containing the mild symptom-associated P domain was joined through the central conserved region to the right half containing the severe symptom-associated V domain. The resulting chimera gave mild symptoms, indicating that the P domain is correlated with changes in pathogenicity of CEV on tomato.[84] The reverse construction will allow further assessment of this interpretation.

Several features in common between the P domains of PSTV, CSV, CEV, TASV, and TPMV are also found in HSV which shows a high sequence homology in this domain (58%) compared to an overall sequence homology of 35% (Table 5). These include a long adenine-dominated oligopurine sequence of 15 to 17 nucleotides on one strand and an oligo (U_{4-7}) sequence on the opposite strand. These features may represent common recognition signals that interact and possibly interfere with homologous host components that lead to symptom expression.

A further possibility is that the P region is involved in the host range of each viroid. For example, the conserved features and sequence homology within the P domain for PSTV, CEV, CSV, HSV, TASV, and TPMV (Table 5) may reflect their overlapping host range of dicotyledonous plants, e.g., tomato, potato, chrysanthemum, and cucumber,[36,142] which is quite distinct from the monocotyledonous palm family host range of CCCV,[143] whose P domain shows low sequence homology with the other viroids (Table 5) and retains only a short conserved adenine sequence of five nucleotides.

4. Sequence Variability of the V Domain

This is the most variable region and shows less than 50% sequence homology between otherwise closely related viroids, such as between TASV and CEV-A, or TPMV and PSTV (Table 5). Similarly, different variants of CEV show considerable variation in this region.[10] It is the V domain, rather than the more highly conserved T2 domain (Table 5) that is responsible for the low sequence conservation reported for the right-half portions of PSTV, CEV, CSV, TASV, and TPMV.[16,32] Sequences in this domain may affect symptom expression. However, the only significant sequence relationship between viroids in the V domain appears to be the presence of an oligo-purine:oligo-pyrimidine helix, usually with a minimum of three G:C pairs.

5. RNA Rearrangements of the Terminal T Domains

The functional role of these domains remains controversial. The termini of some PSTV linear molecules with their 2′,3′-cyclic phosphate have been located in both the T1 and T2 domains,[144,145] suggesting an association with the processing site of viroid replicative intermediates. However, these linear molecules could also have arisen by the nicking of circles in highly susceptible regions during isolation and purification. The T domains are also the preferential binding sites in vitro for purified tomato DNA-dependent RNA polymerase II,[146] but the RNA replicase responsible for the replication of viroids in vivo as well as the site of initiation of RNA synthesis are unknown (see Chapter 2). In addition, other roles for the T domains exist. For example, viroids are presumably transported across membranes to account for their systemic spread and their accumulation, at least for PSTV, in the nucleolus.[147] This movement of rod-like molecules may be achieved by structural signals and orientation of the end loops of which the sequence of the end loop of the T1 domains is highly conserved amongst all PSTV-like viroids (Figure 12) as follows:

UCCUCGGA	PSTV
CCCUCGGG	CEV, CSV, TASV, TPMV
CCCUCUGGG	HSV, CCCV

In addition, sequence data show that three viroids (TASV, TPMV, and CCCV) exhibit unusual relationships with respect to their terminal sequences. For example, TASV shares 73% overall sequence homology with CEV-A but the T2 domains are only 46% homologous (Table 5). In contrast, TASV shares less overall sequence homology with PSTV (64%), but the T2 domains are highly homologous (90%). Therefore, TASV appears to be a recombinant between the T2 domains of a PSTV-ancestral viroid and all but the T2 domain of a CEV-ancestral viroid. TPMV shares 76% overall sequence homology with PSTV but the T1 domains are less homologous (67%). In contrast, TPMV shares less overall sequence homology with CEV-A (60%) but the T1 domains are more homologous (80%). Thus, TPMV appears to be a recombinant between the T1 domain of a CEV-ancestral viroid and all but the T1 domain of a PSTV-ancestral viroid. Finally, CCCV D-O (Figure 5) shares 70% sequence homology with the C domain of PSTV but only 23% sequence homology with respect to the T1 domains. In contrast, CCCV D-O shares low homology with the C domain of HSV (42%), but the T1 domains are more homologous (58%) since the T1 domain of CCCV D-O is almost identical to the left half of the HSV T1 domain (Figure 5). Therefore, CCCV appears to be a recombinant between a viroid with a PSTV lineage and the T1 domain of a HSV-ancestral viroid.

CCCV provides a unique situation in which there are several, specific duplications of the T2 domain.[11] The first RNA species which can be detected after infection of coconut palms with CCCV is the monomeric D-O form of 246 or 247 nucleotides (Figures 1 and 14).[11] As infection progresses, a second larger form of the viroid arises and dominates[67,68] in which the T2 region and part of the V domain are duplicated. Analysis of nine isolates of CCCV from individual palms in the Philippines showed that each isolate contained only one of three possible higher molecular weight forms (D-41, D-50, D-55) in which 41, 50, or 55 nucleotides are duplicated (Figure 14).[11] These duplications are really double duplications of the form X,Y $\rightarrow$ XX,YY, where X = 21, 26, or 28 nucleotides, and Y = 20, 24, or 27 nucleotides. They are found in only three of the nine possible combinations of X and Y: 21 and 20 (CCCC D-41), 26 and 24 (CCCV D-50), and 28 and 27 (CCCV D-55), probably because these three combinations least disrupt the native secondary structure (Figure 14). Since these exchanges occur at different points in the V domain, their occurrence may reflect either structural signals for RNA exchange in the V domain or merely the low functional constraints that exist in this domain, as indicated by the high variability among viroids in this domain (Table 5). A model involving a jumping RNA polymerase has been proposed to explain the origin of these duplications in CCCV during the replication of the viroid.[105]

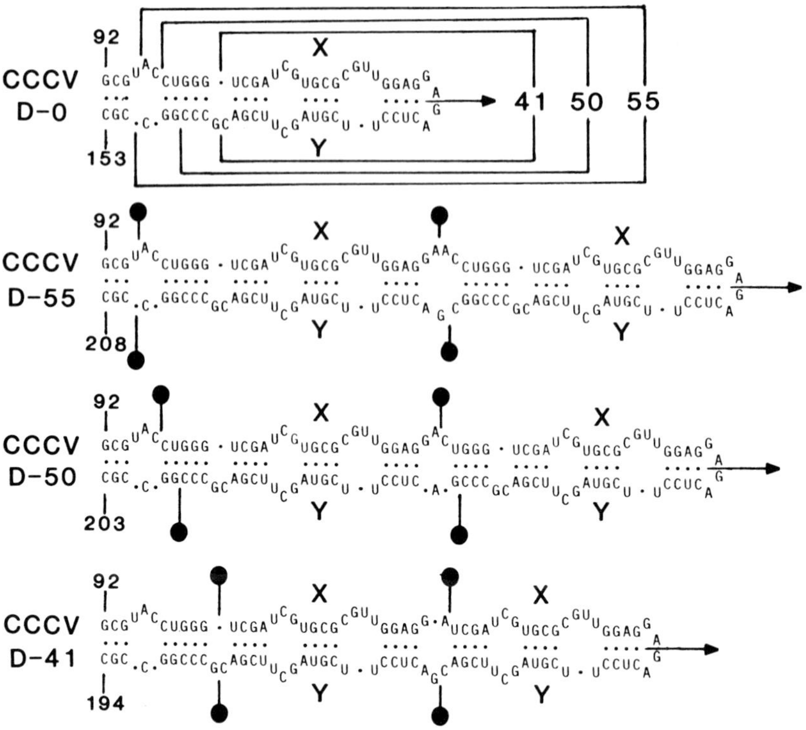

FIGURE 14. Partial sequence duplications of the small form of CCCV (CCCV D-O)[11] in which two adjacent sequences (X,Y) of variable size are repeated, leading to a double duplication of either 41, 50, or 55 nucleotides (D-41, D-50, D-55). The arrows depict the boundary of the X and Y sequences. The filled-in circles mark the boundaries of the duplicated sequences.

C. Structure of Virusoids

1. Sequence and Secondary Structure

The sequences and predicted secondary structures of the four virusoids, as well as that of the satellite RNA of tobacco ringspot virus (sTRSV), are given in Figures 2 and 3. The predicted structures of the virusoids mimic the rod-like structure of viroids. However, for vVTMoV and vSNMV, Steger et al.[96] have proposed a variation in these secondary structures (Figure 2) by an alteration in some of the base pairing and in the inclusion of a bifurcation containing four base pairs between nucleotides 72 and 90.

Despite a similar G:C content to the PSTV-like viroids (Table 3), the virusoids show little sequence homology with them. The only conserved sequence is a GAAAC sequence common to vVTMoV, vSNMV, and all viroids except HSV;[14,40] this sequence occurs in a single-stranded region in the center on the upper strand of the structures in Figures 1 and 2. However, in the case of vLTSV, this GAAAC occurs in a highly base-paired region which is positionally distinct on the lower strand of the predicted structure (nucleotides 210 to 214, Figure 2).[39]

In spite of a difference in size of 11 or 12 nucleotides (Table 2), vVTMoV and SNMV share 94% sequence homology[40] and can really be considered to be sequence variants of the same virusoid. This is similar to the situation with HSV and CPFV, which differ by six nucleotides in size, and the 17 sequence variants of CEV, which vary by five nucleotides.[10]

The isolate of vVTMoV sequenced contained two sequence variants which differed in size by one nucleotide but were otherwise completely homologous.[40] This is a similar situation to CCCV where two sequence variants also differed by only one extra nucleotide.[11]

Table 7
SEQUENCE HOMOLOGY BETWEEN VIRUSOIDS, VIROIDs AND sTRSV

	Sequence homology[a] (%)				
	vVTMoV[b]	v(388)SCMoV	v(332)SCMoV	vLTSV-N	sTRSV
vVTMoV	100				
v(388)SCMoV	41	100			
v(332)SCMoV	42	68	100		
vLTSV-N	35	36	36	100	
sTRSV	39	37	31	40	100
PSTV	27				
ASBV	34				

[a] Initial alignment for the calculation of sequence homology was based on the common GAAAC sequence.
[b] Only vVTMoV is listed, and not vSNMV, since they share 94% overall sequence homology.[40]

The original sequences reported for vVTMoV and vSNMV contained 365/366 and 377 nucleotides, respectively;[40] these sequences were determined by direct RNA sequencing. However, when cDNA clones of vVTMoV and vSNMV were sequenced, an extra C nucleotide was found in both virusoids after nucleotide 48 of the original sequence.[148,149] Hence, their revised lengths are 366/367 and 388 nucleotides, respectively. In the case of vSNMV, the C49 nucleotide contains a stoichiometric amount of 2′-phosphate in addition to the 3′-phosphodiester linkage to nucleotide 50 and the same situation appears to exist for vVTMoV.[149] The presence of this 2′-phosphate may explain why the nucleotide C49 was missed in the original sequencing work as it would have prevented hydrolysis at this linkage by the specific RNases used.[40]

2′-Phosphates may also exist in vLTSV and vSCMoV,[148] but there have been no reports on the presence or absence of a 2′-phosphate in viroids. As suggested by Kiberstis et al.,[149] the presence of a 2′-phosphate in the encapsidated virusoids, as compared to its lack in other host RNAs where it is rapidly removed in the normal synthetic pathways, may be related to the rapid protection from phosphatases offered by packaging into virus particles. Kiberstis et al.[149] have also suggested that the presence of a 2′-phosphate at a unique site in vSNMV marks the site of ligation of linear vSNMV, produced during the replication process, to the final circular form. 2′-Phosphomonester,3′-5′-phosphodiester bonds are known to result from one type of ligation event.[150] Further details on this type of reaction are contained in Chapter 2.

2. Sequence Homology Between Virusoids and sTRSV

Except for v(388)SCMoV and v(332)SCMoV which show 68% sequence homology (Table 7), any two virusoid species show a sequence homology in the range of 35 to 42% (Table 7). A comparison of sTRSV with the virusoids also shows similar sequence homology (31 to 40%), values only a little higher than the sequence homology between vVTMoV and a G:C-rich viroid such as PSTV (27%, Table 7). A most interesting 50 nucleotide region of strong sequence homology among vLTSV, vVTMoV, vSNMV, the two vSCMoV species, and sTRSV is shown in Figure 15.[55-57] Indeed, vVTMoV and vSNMV show greater sequence homology in this region with sTRSV, whose helper virus belongs to the Nepovirus group, than to the other virusoids whose helper viruses all belong to the Sobemovirus group.[50] Since the sequence alignments include the juxtaposition of the 3′- and 5′-termini of the sTRSV linears, it was proposed that this indicated the site of processing of longer than unit length forms of sTRSV and the virusoids during the replication cycle to give unit-length

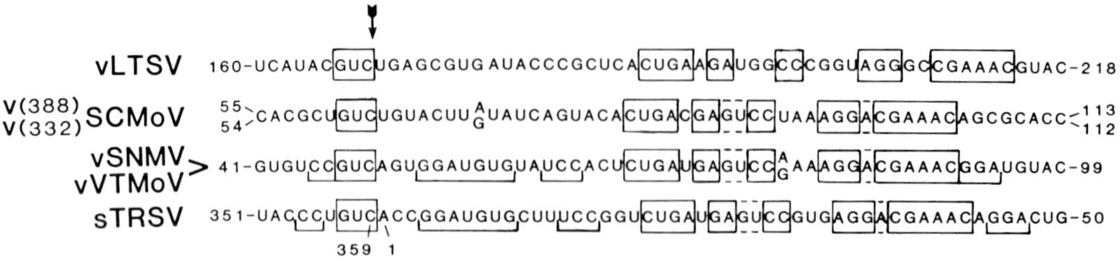

FIGURE 15. Region of high sequence homology found in all virusoids and shared with sTRSV.[41,56,57] The solid boxes indicate sequence homology between the five virusoids and sTRSV; dashed boxes indicate the sequence homology between sTRSV and all virusoids except vLTSV; brackets indicate sequence homology between sTRSV, vSNMV and vVTMoV. The arrow indicates a postulated processing site of viroid precursor RNAs that corresponds to the termini of sTRSV.[56,57]

linear molecules.[56,57] In the case of virusoids, these would then be ligated to the dominant circular form found in vivo.[56,57] This prediction now seems likely in view of the detection of a 2′-phosphate at the proposed processing site.[149]

3. RNA Rearrangements of Two Virusoids of SCMoV

Different isolates of SCMoV contain either a large virusoid, v(388)SCMoV, of 388 nucleotides, and/or a smaller virusoid, v(332)SCMoV, of 332 nucleotides.[41,48] The same helper virus appears to be present in all SCMoV isolates since restriction enzyme digests made from SCMoV cDNA gave identical patterns on PAGE for all isolates studied.[41,56] The left-hand 218 nucleotides of each virusoid show 95% sequence homology but the remaining nucleotides of each molecule differ markedly in size and sequence, showing only 25% sequence homology[57] (Figure 16). We have postulated that these sequence differences have arisen by RNA exchange with as yet undiscovered RNA molecules in an analogous way to the T domain rearrangements proposed for PSTV-like viroids.[105]

4. Interaction of Virusoids with Helper Viruses

Virusoids are dependent upon a helper virus for their replication and they are encapsidated by the coat protein coded for by that virus.[1,2] These interactions of the virusoid with functions provided by the helper virus presumably rely on as yet undiscovered sequence and structural signals in the virusoid. Although specific combinations of virusoid and helper virus were isolated in the field, at least one helper virus, LTSV, will support the replication of a number of different virusoids such as vLTSV, v(388)SCMoV, and vSNMV.[39,52,151] This suggests that these three virusoids may share a common recognition signal which allows for the interaction with the helper LTSV.

The most highly conserved region among virusoids appears to be around the putative processing site for virusoid precursors (Figure 15). Outside this region, the longest conserved sequence is a GAUUUU[39] which is not conserved in sTRSV whose helper virus belongs to a different virus group than that of virusoids. However, the GAUUUU cannot be the only signal necessary for recognition since SNMV will not support the replication of vLTSV.[52] The latter observation is surprising since two other viruses in the Sobemovirus group which have not been found associated naturally with a virusoid (southern bean mosaic virus and sowbane mosaic virus) will, nevertheless, support the replication of vLTSV.[152,153] The complete sequencing of the RNAs of viruses which support virusoid replication will be important in further analysis of this situation.

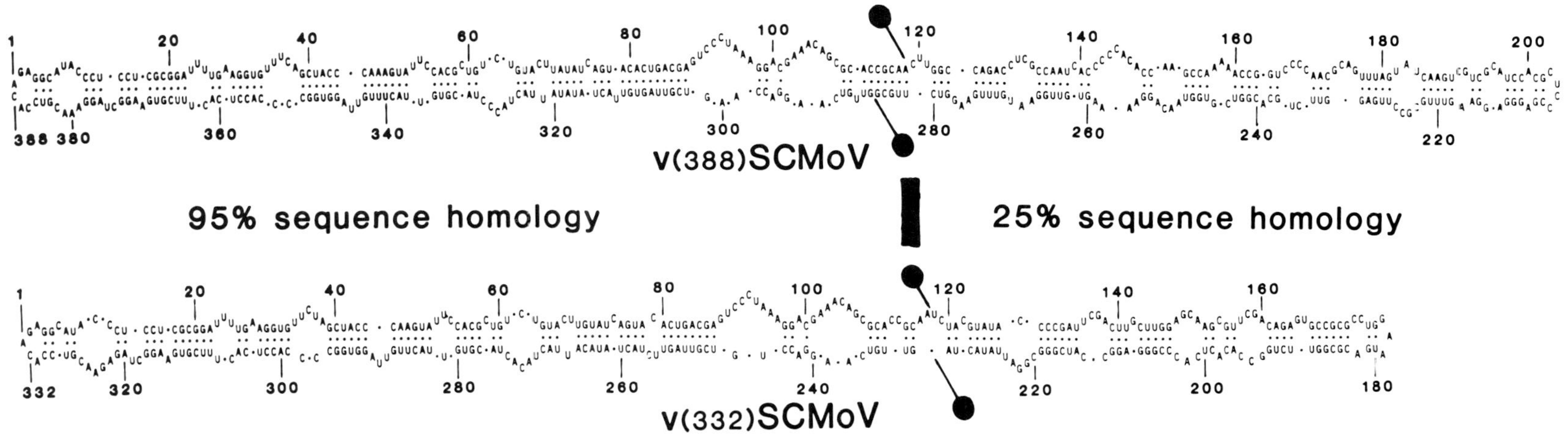

FIGURE 16. Regions of high and low sequence homology between the two virusoids of SCMoV.[4] The filled-in circles mark the boundaries.

V. STRUCTURE AS RELATED TO BIOLOGICAL ACTIVITY

A. There is No Evidence for In Vivo Translation Products of Viroids and Virusoids

Although several lines of indirect evidence considered below indicate that viroids and virusoids lack mRNA activity, the possibility that they can code for small amounts of low molecular weight proteins cannot be completely discounted. It is considered prudent to keep an open mind until more definitive evidence becomes available. It is feasible that low molecular weight proteins may have a role in controlling gene expression in eukaryotes. For example, Hillar et al.[154] described M_r 600 to 1500 proteins which are associated with DNA and nuclear RNA and which inhibit transcription and translation in vitro. Evidence arguing against viroids and virusoids coding for functional polypeptides, however, is as follows:

1. PSTV and CEV are not translated in vitro in systems derived from bacteria, wheat, and other eukaryote systems.[155-157]
2. Eukaryotic ribosomes do not bind to circular RNA, an essential step for the initiation of translation.[158] However, the possibility that subgenomic linear fragments of viroids or virusoids can be tranlated has never been tested. The ability to produce significant quantities of defined, linear RNA transcripts from cDNA clones of viroids and virusoids in appropriate vectors[10,87,88] makes such experiments feasible.
3. Low molecular weight proteins specific for infection by PSTV or CEV have not been detected in vivo.[159,160] However, the methods used in this work were not suitable for detecting low levels of viroid-induced proteins.
4. PSTV and CCCV D-O do not contain an AUG initiation codon in either the viroid or its complementary strand,[11,15] which is a necessary prerequisite for translation by eukaryotic ribosomes.[161] Although one AUG-initiated polypeptide is possible in the minus strand of CEV-A,[8] potential translation products are not conserved among 17 sequence variants of CEV.[10] In addition, there is a lack of conservation between possible translation products of both the plus and minus strands of the closely related viroids, CEV and CSV.[5,8] These data provide strong evidence that viroids lack mRNA activity.
5. PSTV accumulates in the nucleolus[147] and is not found associated with polyribosomes, or in chloroplasts or mitochrondria where initiation codons may operate. The distribution in the cell of viroids such as CEV, CSV, and ASBV, which do contain AUG codons, has yet to be determined.
6. vSNMV, vVTMoV, and vLTSV contain from three to six AUG codons in their plus and minus strands. However, there is no conservation of potential AUG-initiated translation products between vSNMV and the 365 and 366 nucleotide forms of vVTMoV,[40] despite 94% sequence homology. Even for the two closely related sequence variants of vLTSV, only two of the seven potential AUG-initiated polypeptides are conserved.[39] Hence, the virusoids also seem to lack mRNA activity in vivo.

B. Correlation of Viroid Sequences to Symptom Expression

PSTV, CEV, CSV, TASV, TPMV, and HSV (CPFV) are all able to infect tomato but with variable degrees of virulence both among viroid species as well as among sequence variants of the individual viroid species.[10,16,142] So far, sequence changes in the P domain (Table 6) of PSTV[16] and the P and V domains of CEV have been correlated with symptom expression. However, it is unknown what signal these sequence changes provide.

Sänger[16] has noted that six sequence variants of PSTV that elicit symptoms in tomato ranging from mild, intermediate, and severe to lethal, correspond to an increase of $\triangle G$ in the P domain. It is to this decrease of stability in the secondary structure of the P domain that Sänger has attributed an increase in severity of symptoms. However, the opposite is

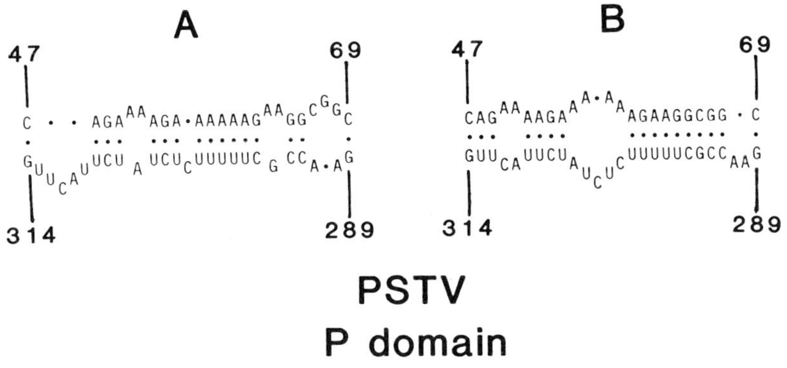

FIGURE 17. Alternative structures proposed for PSTV[15] in the P domain. (A) Secondary structure based on susceptibility to cleavage by single-strand specific nucleases. (B) Secondary structure based on theoretically derived thermodynamic principles established by Tinoco et al.[94]

the case for the sequence variants of CEV,[10] so that other properties of this domain must be involved in pathogenicity.

Part of the difficulty in associating conformational changes with variation in symptom expression can be accounted for by the uncertainity of the native secondary structure in the P domain of PSTV. The conformation (Figure 17A) that best accommodates sites sensitive to single-strand specific RNases or to the ability of cytosine nucleotides to be modified by bisulfite treatment differs from the theoretically most stable form (lowest ΔG) of the P domain of PSTV (Figure 17B).[15] In addition, there is the possibility of local tertiary structures in the P domain about which we have no data, but which could be similar to the tertiary structure discovered in the conserved central core of PSTV[110] (see Section III.C).

Another difficulty in correlating sequence changes in the P domain with pathogenicity concerns the role of long-range interactions between different parts of the viroid molecule in affecting pathogenicity. For example, the ability of PSTV-like viroids to form stem II as proposed from thermodynamic and electron microscopic data (Figure 6; Section III.C.1) may be affected by neighboring sequence changes in the P or V domains. The only evidence of such long-range interactions playing a role in viroid function comes from a chimera of the left half of CSV joined to the right half of CEV,[84] which was not infectious. This chimera maintains a conserved stem I (Figure 7) but does not allow the formation of stem II. Other factors, however, may explain the noninfectivity of the chimera.

C. Sequence Homology Between the Putative Processing Sites of ASBV, sTRSV, and the Virusoids

Multimeric plus and minus RNA forms of sTRSV, vVTMoV, vSNMV, and vLTSV have been detected in virions and in nucleic acid preparations of infected plants.[57,58,60,61,63] The RNA species may represent products of some form of rolling circle replication similar to viroids[61,62] (see Chapter 2).

The processing of these greater-than-unit-length RNA forms to the monomeric RNA form has been examined in vitro for dimers of sTRSV[162] and ASBV.[69,89] In the case of sTRSV, single-stranded dimers were isolated from virions and could be cleaved nonenzymatically with up to 70% efficiency in the presence of a divalent cation resulting in a 5′-hydroxyl and a 2′,3′-cyclic phosphate at the termini. The sequences at the termini were the same as those at the ends of naturally occurring linear molecules of sTRSV isolated from virions.[55,162] With ASBV, dimeric RNA transcripts of both orientations of ASBV were prepared from dimeric clones of ASBV cDNA in the pSP6 vector[87] and these processed rapidly in the reaction mixture to give linear plus and minus monomers with a 5′-hydroxyl and a 2′,3′-cyclic phosphate.[89] Evidence that this processing was nonenzymic was obtained by purification by PAGE of the residual full-length dimers plus and minus transcripts which then processed to monomers in the presence of only Mg^{2+} or Ca^{2+} and Tris-HCl, pH 7 to 9.[89,163] RNA transcripts from partial length plus and minus clones of vLTSV-A also self-processed at unique sites in the plus and minus RNA[164] under similar conditions to ASBV.

The functional similarity of the nonenzymic processing in vitro of plus and minus ASBV, sTRSV, and plus and minus vLTSV-A extends to homology in sequence and possible structure surrounding the self-processing sites of both RNAs (Figure 18). In the case of ASBV and vLTSV-A, the numbers of nucleotides in the structures postulated to signal processing amount to about one third of each molecule (Figures 2 and 18). Homologous sequences and structures can also be found in the region of the putative processing site of the plus RNAs of other virusoids (Figures 5 and 18). This sequence and structural homology also indicates that ASBV may provide an evolutionary link between viroids and virusoids. Alternatively, the conserved region between ASBV, sTRSV, and the virusoids may reflect convergent evolution of function, since they share little sequence homology elsewhere in their respective molecules.

D. Future Considerations

1. In Vitro Mutagenesis to Investigate Viroid and Virusoid Function

The powerful approach of in vitro mutagenesis has become possible at the DNA level since viroid cDNA clones and/or RNA transcripts from them are infectious.[112,135,136,165] On the basis of our extensive knowledge of viroid sequences, we can now alter individual nucleotides to test a number of models of viroid function. Mutagenesis has already been used to discriminate between regions of CEV that are responsible for changes in symptom expression by constructing a chimera from mild and severe variants of CEV.[84] Site-directed point mutations[166] can be used to test for nucleotides crucial to successful infection as well as changes in symptom expression and host range. For example, a point mutation of G → U at either nucleotide 351[84] or nucleotide 97[112] of CEV abolishes infectivity. Deletion mutagenesis[167] has been used to examine essential sequences required for the infectivity of HSV cDNA clones.[139] Chimeric constructions between different viroids can be used to test for compatibility and the possibility of generating new viroid species. The same procedures that are being used for viroids are directly applicable to virusoids. In addition, there is the interesting aspect of investigating the specificity of the relationship between the virusoid and its helper viral RNA.

Finally, with the growing importance of mutagenesis, other variations may be employed to test structure-function relationships. For example, mutagenesis of noninfectious mutants back to infectious forms with generalized mutagenic agents such as nitrous acid, nitrosoguanidine, or X-rays may provide viable sequence variants with compensatory changes in addition to any revertants.

2. Use of Viroids as Plant Vectors

Viroid reconstructions at the DNA level, together with the ability of viroids to spread systemically in a wide range of hosts, often in the absence of symptom expression, open up the possibility for the use of viroids as vectors for the introduction of new genetic material into plants. However, a few words of caution are necessary. Viroids are particularly noted for their small size (246 to 375 nucleotides) which may be of functional significance. Changes in the form of point mutations or four nucleotide deletions have abolished infectivity of CEV,[84,112] suggesting that relatively few changes may be acceptable without loss of infectivity. Even if the successful introduction of foreign genetic material can be achieved, its maintenance may prove difficult because of the rapid rate of mutations during RNA-directed

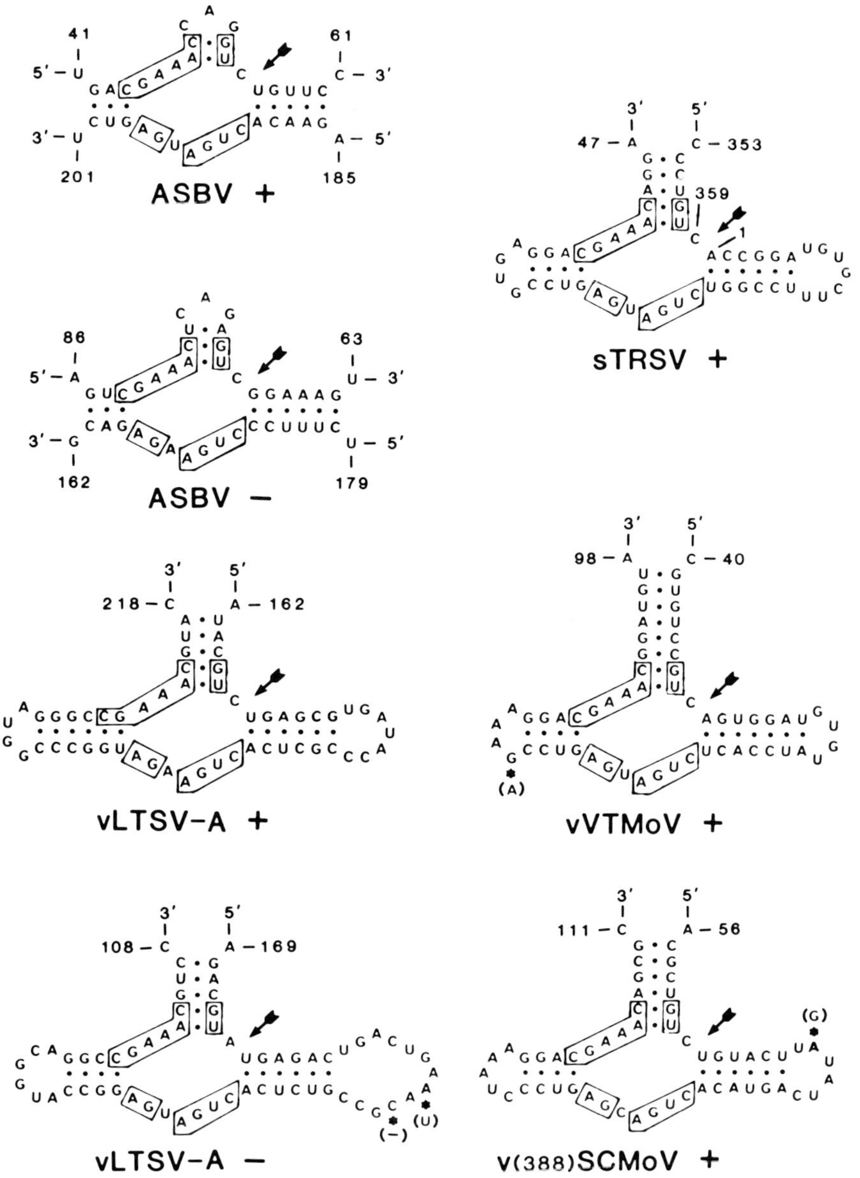

FIGURE 18. Sequence homologies (boxed) and structural homologies adjacent to the in vitro nonenzymic processing sites (arrows) of ASBV plus and minus RNA, sTRSV plus RNA, vLTSV plus and minus RNA, and the postulated processing sites (arrows) in vVTMoV and v(388)SCMoV.[89,164] The asterisks mark nucleotide differences between vLSTV-A and vLSTV-N, vVTMoV and vSNMV, and v(388)SCMoV and v(332)SCMoV. Sequence numbers of minus RNAs are assigned according to the nucleotide numbers given in the complementary plus RNA. Homologous sequences and structures are not apparent in the minus RNA sequences of vVTMoV, vSNMV, v(388)SCMoV, v(332)SCMoV, and sTRSV.

transcription[168] and the potential for RNA rearrangements to regenerate the original viroid. Nevertheless, partial viroid sequences may prove useful as part of larger DNA vector constructs to act either as origins of replication or transcription if their signals are retained at the DNA level, or as a means of amplifying vector RNA transcripts.

VI. POSSIBLE ORIGINS OF VIROIDS

A. Diversity of Predicted Origins

Viroids have been considered many times before to be escaped regulatory elements originating from prokaryotes, viral RNA, or plant host RNA.[6,17,19,30,169-174] Such a diversity of affinities has been postulated due to the many unusual characteristics of viroids together with the paucity of knowledge regarding viroid function. For example, the in vivo RNA replicase of viroids is unknown, yet a number of purified prokaryotic, viral, and host RNA polymerases are capable of transcribing viroids[57] (see Chapter 2). In addition, viroids show striking, but incomplete, sequence and structural homologies with different viral and host RNAs.

B. Sequence and Structural Affinities of Viroids with RNA Viruses

Although viroids possess several features that clearly distinguish them from viruses (see Section I), two viral groups, retroviral proviruses and plant viral satellite RNAs, have been examined for possible relationships with viroids. For example, it has been postulated that several features of the C domain of PSTV, TPMV, CEV, TASV, and CSV (but not CCCV and HSV) resemble those of transposable genetic elements, in particular retroviral proviruses.[17] These include an imperfect (5/9 bases) direct repeat bordered by CA and UG dinucleotides characteristic of all retroviral proviruses, and which forms part of a nearly complete inverted repeat (6/7 bases). In addition, these viroids possess a polypurine sequence (11 to 18 bases) similar to the putative reverse transcriptase primer sequence of retroviral proviruses. On the basis of these characteristics, Kiefer et al.[17] proposed that viroids may have originated from transposable genetic elements, such as retroviral proviruses, by deletion of all protein encoding genes and most of the long terminal repeats and utilization of the polypurine sequence for RNA synthesis rather than for signaling RNA-directed DNA synthesis.

As considered in detail in this review, the virusoids represent a second class of viral RNAs with viroid-like characteristics. Although there is very limited sequence homology between viroids and virusoids and significant differences in thermo- and hydrodynamic properties, there are a few pertinent features in common. These include low M_r circular single-stranded RNA, presumed lack of mRNA activity, a potentially similar rolling circle pathway for replication involving greater-than-unit-length plus and minus strands,[57,61,63] an unusual sequence relationship between the two virusoids of SCMoV which mimics the apparent chimeric nature of certain viroids, and homologous sequences surrounding the putative processing sites of virusoids and ASBV (Figure 18).

Recently, a linear satellite RNA of peanut stunt virus (sPSV) has been shown to share more regions of high sequence homology with viroids, in particular CSV, than with the related linear satellites associated with cucumber mosaic virus.[175] sPSV also shares sequences in common with both strands of the C domain conserved among all PSTV-like viroids except HSV, including a GAAAC sequence also present in ASBV and virusoids. However, this sequence in sPSV cannot form similar conserved structures that are feasible among viroids (Figures 7 and 13) or virusoids and ASBV (Figure 18). It remains to be determined if these sequences signal the same function(s) in sPSV as in viroids and virusoids where they may be associated with the processing of replicative intermediates. Some differences might be expected since linear forms of viroids and virusoids terminate with a 5′-hydroxyl and a 2′,3′-cyclic phosphate, whereas sPSV has a 5′-m^7G cap and 3′-hydroxyl.[175]

Any speculations on evolutionary relationships arising from such comparisons are difficult to test experimentally. The results of these comparisons, however, may eventually provide significant guidelines in designing experiments to correlate structure with function.

C. Sequence and Structural Homologies of Viroids with Host RNAs

Sequence homology has been noted between small nuclear (sn) RNAs of animals and viroids, in particular Novikoff hepatoma U5a RNA and ASBV,[176] Novikoff hepatoma U3B RNA and PSTV,[147,177] and U1 RNA with all viroids except ASBV.[6,11,14,17,172,173] The assumption in all cases is that plant equivalents to animal snRNAs are present. However, in one case where U5 RNA from pea nuclei has been completely sequenced, only 35% sequence homology with vertebrate U5 RNAs was found.[178] Until other plant snRNAs are sequenced, it is difficult to make a reasonable assessment of the meaning of any homology between viroids and snRNAs.

Another proposal is that viroids have evolved from intervening sequences by virtue of sequence homology with the 5′-end of animal U1 RNA. This region appears to be responsible for the excision of introns from pre-mRNA by base pairing to both ends of the intron.[179] It has been proposed that the complementary sequence of viroids (except ASBV) could base pair with U1 RNA in a manner similar to introns, reflecting a past event that led to the appearance of a viroid-like RNA from a host component.[172]

A variation on the escaped intron hypothesis is based on the localization of PSTV in the nucleolus of tomato cells[147] where precursor ribosomal RNA, rather than pre-mRNA, is processed. Sequence and structural homology has been reported to occur between viroids and the intron of *Tetrahymena* nuclear rRNA which is capable of self-induced processing and circularization.[180,181] It remains to be established, however, whether the processing signals for ASBV and other viroid precursor replication intermediates are homologous to any previously published RNA processing mechanisms.[182]

The notion of viroids as escaped host regulatory elements generated from the eukaryotic splicing machinery has been extended by Zimmern,[174] who has proposed a scheme whereby the development and spatial arrangement of cells in eukaryotes are controlled by exchange of genetic information between cells. This genetic transfer is postulated to be mediated by incomplete RNA elements, "signal RNAs", capable of initiating RNA-directed amplification. By a reversal of the splicing process, these signal RNAs could attach to "antenna RNAs" forming the functionally complete entities, "fusons". Viroids could represent examples of aberrant signal RNAs or modified fusons.

D. RNA Rearrangements in the Evolution of Viroids

Although all the hypotheses considered earlier envisage the involvement of some form of splicing to account for the origin of viroids, a different mechanism of viroid evolution has been postulated involving discontinuous transcription.[105] This scheme was derived by analogy with the evolution of defective interfering (DI) particles of animal viruses[183] and used to explain the unusual sequence relationships between the T domains of certain viroids and the distinctive partial duplications of CCCV (Section IV.B.5). It is feasible that such events could lead to new viroids with the ability to infect new hosts. The outbreak and subsequent apparent spread of the cadang cadang disease of coconuts in the Philippines in this century[24] are consistent with such an origin for CCCV.

Although intermolecular RNA rearrangements giving rise to new chimeric viroid species remain hypothetical, other apparently chimeric RNA virusoids and viruses are known. As already considered, the two virusoids of SCMoV contain a left-hand end of 218 nucleotides with 95% sequence homology but the rest of each molecule varies in size and shows only 25% sequence homology. Another example is that of bovine leukemia virus, a retrovirus whose *pol* gene product shows strong amino acid sequence homology with the *pol* gene product of Rous sarcoma virus but whose *env* gene product is more homologous to Maloney murine leukemia virus.[184] Finally, amino acid sequence homologies, noted among nonstructural proteins encoded by animal and plant RNA viruses with different modes of gene organization and expression, suggest that RNA rearrangements leading to reassortment of

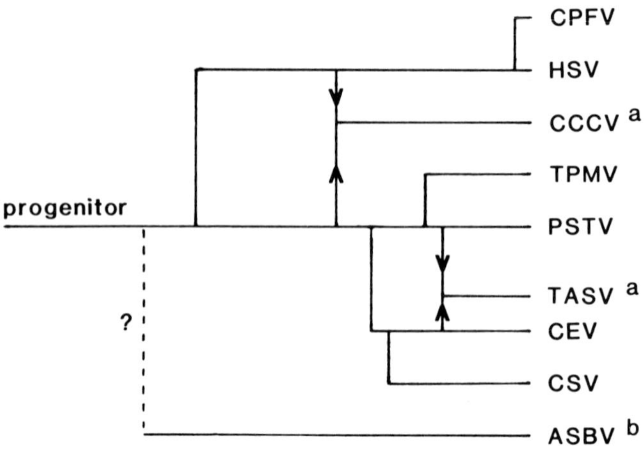

FIGURE 19. Possible evolutionary tree for viroids based on sequence homology between viroids. (a) In this scheme, CCCV and TASV are considered to have arisen by RNA exchange between two separate viroid species as indicated by arrows. (b) ASBV shows low sequence homology with other viroids and may indicate a polyphyletic origin for viroids.

functional genetic and protein domains may be an important mechanism in RNA virus evolution.[185]

In summary, intermolecular RNA rearrangements in viroids present the possibility of an intriguing evolutionary tree (Figure 19) in which RNA exchange between distinct viroid species could give rise to new infectious RNA molecules. Superimposed is the chance of RNA exchange of viroids with both cellular and viral RNA components.

ACKNOWLEDGMENTS

The work from our laboratory, which is described in this review, was supported by the Australian Research Grants Scheme, the Rural Credits Development Fund of the Reserve Bank of Australia, and by a Commonwealth Government Grant to the Adelaide University Centre for Gene Technology in the Department of Biochemistry. We thank all members of our research group for unpublished data, A. C. Forster and J. E. Visvader for discussions, G. Bruening, J. Haseloff, T. Meshi, M. W. Schwinghamer, and Lita Imperial for access to unpublished data and discussions, Jennifer Cassady for assistance with the figures, and Leanne Goodwin and Jane Somerville for typing the manuscript.

GLOSSARY

Helper virus. A virus capable of supporting the replication of an additional, nonessential, and unrelated nucleic acid component; e.g. a virusoid or satellite RNA.

Isolate of viroid or virusoid. Viroid or virusoid isolated from a single infected plant. May consist of only one, or more than one, sequence variant.

Satellite RNA. An encapsidated RNA component which is dependent upon a helper virus for replication but shows no significant sequence homology with either the viral or host genomes. Satellite RNAs can be classed into a number of different groups depending on their structural and biological properties.[1,2] The circular virusoids form one of these groups. Many satellite RNAs fall within the range of 300 to 400 nucleotides and are dependent upon the helper virus for encapsidation. Larger satellite RNAs encode their own coat protein; such encapsidated RNAs are called satellite viruses.

Sequence homology. Calculated as the total number of matching nucleotides between two optimally aligned sequences in which each matching subsection consists of a minimum of three identical nucleotides, divided by the total number of nucleotides used for sequence comparisons, and expressed as a percent.

Sequence variant of viroid or virusoid. An individual RNA component. Isolates of viroids or virusoids may contain a mixture of sequence variants where each variant differs by one or more nucleotides.

Viral genome. The minimum genetic material of a virus required for successful infection and maintenance under field conditions.

Viroid. Molecular parasite of higher plants composed of naked, single-stranded, low molecular weight, circular RNA (M_r 0.8 to 1.3 × 10^5) which utilizes only host components for its replication. There is no evidence that any viroid can code for one or more polypeptides in vivo. They exist in solution as rod-like structures arranged in a series of short base-paired and nonbase-paired regions.

Viriod species. A population of one or more independently replicating sequence variants which show more than 90% sequence homology. Cucumber pale fruit viroid and hop stunt viroid show greater than 90% sequence homology and are therefore sequence variants of the same viroid species. Citrus exocortis viroid is a viroid species containing at least 17 naturally occurring sequence variants.

Virusoid. Single-stranded, circular, low molecular weight, viroid-like RNA (M_r 1.1 to 1.3 × 10^5). These satellite RNAs are dependent on a helper RNA virus for replication and are encapsidated in the circular form by that virus. They exist in solution as rod-like structures with secondary structures similar to viroids.

REFERENCES

1. **Murant, A. F. and Mayo, M. A.,** Satellites of plant viruses, *Ann. Rev. Phytopathol.*, 20, 49, 1984.
2. **Francki, R. I. B.,** Plant virus satellites, *Ann. Rev. Microbiol.*, 39, 151, 1985.
3. **Palukaitis, P., Hatta, T., Alexander, D. McE., and Symons, R. H.,** Characterization of a viroid associated with avocado sunblotch disease, *Virology*, 99, 145, 1979.
4. **Symons, R. H.,** Avocado sunblotch viroid: primary sequence and proposed secondary structure, *Nucleic Acids Res.*, 9, 6527, 1981.
5. **Haseloff, J. and Symons, R. H.,** Chrysanthemum stunt viroid: primary sequence and secondary structure, *Nucleic Acids Res.*, 9, 2741, 1981.
6. **Gross, H. J., Krupp, G., Domdey, H., Raba, M., Jank, P., Lossow, C., Alberty, H., Ramm, K., and Sänger, H. L.,** Nucleotide sequence and secondary structure of citrus exocortis and chrysanthemum stunt viroid, *Eur. J. Biochem.*, 121, 249, 1982.
7. **Schlemmer, A., Roistacher, C. N., and Semancik, J. S.,** Detection of a unique, infectious RNA which causes symptoms in citron typical of citrus exocortis disease, *Phytopathology*, 75, 946, 1985.
8. **Visvader, J. E., Gould, A. R., Bruening, G. E., and Symons, R. H.,** Citrus exocortis viroid: nucleotide sequence and secondary structure of an Australian isolate, *FEBS Lett.*, 137, 288, 1982.
9. **Visvader, J. E. and Symons, R. H.,** Comparative sequence and structure of different isolates of citrus exocortis viroid, *Virology*, 130, 232, 1983.
10. **Visvader, J. E. and Symons, R. H.,** Eleven new sequence variants of citrus exocortis viroid and the correlation of sequence with pathogenicity, *Nucleic Acids Res.*, 13, 2907, 1985.
11. **Haseloff, J., Mohamed, N. A., and Symons, R. H.,** Viroid RNAs of cadang-cadang disease of coconuts, *Nature (London)*, 299, 316, 1982.
12. **Owens, R. A., Smith, D. R., and Diener, T. O.,** Measurement of viroid sequence homology by hybridization with complementary DNA prepared *in vitro*, *Virology*, 89, 388, 1978.
13. **Sano, T., Uyeda, I., Shikata, E., Ohno, T., and Okada, Y.,** Nucleotide sequence of cucumber pale fruit viroid: homology to hop stunt viroid, *Nucleic Acids Res.*, 12, 2437, 1984.
14. **Ohno, T., Takamatsu, N., Meshi, T., and Okada, Y.,** Hop stunt viroid: molecular cloning and nucleotide sequence of the complete cDNA copy, *Nucleic Acids Res.*, 11, 6185, 1983.

15. **Gross, H. J., Domdey, H., Lossow, C., Jank, P., Raba, M., Alberty, H., and Sänger, H. L.,** Nucleotide sequence and secondary structure of potato spindle tuber viroid, *Nature (London),* 273, 203, 1978.
16. **Sänger, H. L.,** Minimal infectious agents: the viroids, in *The Microbe 1984: Part I Viruses,* Mahy, B. W. J. and Pattison, J. R., Eds., Society for General Microbiology, Cambridge University Press, London. 1984, 281.
17. **Kiefer, M. C., Owens, R. A., and Diener, T. O.,** Structural similarities between viroids and transposable genetic elements, *Proc. Natl. Acad. Sci. U.S.A.,* 80, 6234, 1983.
18. **Chen, W., Tien, P., Zhu, Y. X., and Liu, Y.,** Viroid-like RNAs associated with burdock stunt disease, *J. Gen. Virol.,* 64, 409, 1983.
19. **Diener, T. O.,** *Viroids and Viroid Diseases,* John Wiley & Sons, New York, 1979.
20. **Boccardo, G., Beaver, R. G., Randles, J. W., and Imperial, J.,** Tinangaja and bristle top, coconut diseases of uncertain etiology in Guam, and their relationship to cadang-cadang disease of coconut in the Phillippines, *Phytopathology,* 71, 1104, 1981.
21. **Keese, M., Hutchins, C. J., Keese, P., and Symons, R. H.,** Unpublished data.
22. **Diener, T. O.,** Potato spindle tuber "virus". IV. A replicating, low molecular weight RNA, *Virology,* 45, 411, 1971.
23. **Hollings, M. and Stone, O. M.,** Attempts to eliminate chrysanthemum stunt from chrysanthemum by meristem-tip culture after heat-treatment, *Ann. Appl. Biol.,* 65, 311, 1970.
24. **Zelazny, B., Randles, J. W., Boccardo, G., and Imperial, J. S.,** The viroid nature of the cadang-cadang disease of coconut palm. *Sci. Filip.,* 2, 45, 1982.
25. **Semancik, J. S. and Weathers, L. G.,** Exocortis virus: an infectious free-nucleic acid plant virus with unusual properties, *Virology,* 47, 456, 1972.
26. **Sänger, H. L.,** An infectious and replicating RNA of low molecular weight: the agent of the exocortis disease of citrus, *Adv. Biosci.,* 8, 103, 1972.
27. **Diener, T. O. and Lawson, R. H.,** Chrysanthemum stunt: a viroid disease, *Virology,* 51, 94, 1973.
28. **Hollings, M. and Stone, O. M.,** Some properties of chrysanthemum stunt, a virus with the characteristics of an uncoated ribonucleic acid, *Ann. Appl. Biol.,* 74, 333, 1973.
29. **Riesner, D., Henco, K., Rokohl, U., Klotz, G., Kleinschmidt, A. K., Domdey, H., Jank, P., Gross, H. J. and Sänger, H. L.,** Structure and structure formation of viroids, *J. Mol. Biol.,* 133, 85, 1979.
30. **Sänger, H. L.,** Biology, structure, functions and possible origin of viroids, in *Encyclopedia of Plant Physiology New Series,* Vol. 14B, Parthier B. and Boulter, D., Eds., Springer-Verlag, Berlin, 1982, 368.
31. **Diener, T. O.,** Viroids, *Adv. Virus Res.,* 28, 241, 1983.
32. **Riesner, D. and Gross, H. J.,** Viroids, *Ann. Rev. Biochem.,* 54, 531, 1985.
33. **Howell, S. H.,** Physical structure and genetic organization of the genome of maize streak virus (Kenyan isolate), *Nucleic Acids Res.,* 12, 7359, 1984.
34. **Mullineaux, P. M., Donson, J., Morris-Krsinich, B. A. M., Boulton, M. I., and Davis, J. W.,** The nucleotide sequence of maize streak virus DNA, *EMBO J.,* 3, 3063, 1984.
35. **Fiers, W., Contreras, R., Duerinck, F., Haegeman, G., Iserentant, D., Merregaert, J., Min Jou, W., Molemans, F., Raeymaekers, A., Van den Berghe, A., Volckaert, G., and Ysebaert, M.,** Complete nucleotide sequence of bacteriophage MS2 RNA: primary and secondary structure of the replicase gene, *Nature (London),* 260, 500, 1976.
36. **Singh, R. P.,** Viroids and their potential danger to potatoes in hot climates, *Can. Plant Dis. Sur.,* 63, 13, 1983.
37. **Da Graca, J. V. and Van Vuuren, S. P.,** Use of high temperature to increase the rate of avocado sunblotch symptom development in indicator seedlings, *Plant Dis.,* 65, 46, 1981.
38. **Riesner, D., Steger, G., Schumacher, J., Gross, H. J., Randles, J. W., and Sänger, H. L.,** Structure and function of viroids, *Biophys. Struct. Mech.,* 9, 145, 1983.
39. **Keese, P., Bruening, G., and Symons, R. H.,** Comparative sequence and structure of circular RNAs from two isolates of lucerne transient streak virus, *FEBS Lett.,* 159, 185, 1983.
40. **Haseloff, J. and Symons, R. H.,** Comparative sequence and structure of viroid-like RNAs of two plant viruses, *Nucleic Acids Res.,* 10, 3681, 1982.
41. **Haseloff, J., Davis, C., and Symons, R. H.,** Unpublished data.
42. **Randles, J. W., Davies, C., Hatta, T., Gould, A. R., and Francki, R. I. B.,** Studies on encapsidated viroid-like RNA. I. Characterization of velvet tobacco mottle virus, *Virology,* 108, 111, 1981.
43. **Greber, R. S.,** Some characteristics of solanum nodiflorum mottle virus — a beetle-transmitted isometric virus from Australia, *Aust. J. Biol. Sci.,* 34, 369, 1981.
44. **Gould, A. R. and Hatta, T.,** Studies on encapsidated viroid-like RNA. III. Comparative studies on RNAs isolated from velvet tobacco mottle virus and *Solanum nodiflorum* mottle virus, *Virology,* 109, 137, 1981.
45. **Tien-Po, Davies, C., Hatta, T., and Francki, R. I. B.,** Viroid-like RNA encapsidated in lucerne transient streak virus, *FEBS Lett.,* 132, 353, 1981.
46. **Forster, R. L. S. and Jones, A. T.,** Properties of lucerne transient streak virus, and evidence of its affinity to southern bean mosiac virus, *Ann. Appl. Biol.,* 93, 181, 1979.

47. **Greber, R. S.,** Solanum nodiflorum virus, *Glasshouse Crops Res. Inst. Rep.*, 150, 1978.
48. **Francki, R. I. B., Randles, J. W., Hatta, T., Davies, C., and Chu, P. W. G.,** Subterranean clover mottle virus: another virus from Australia with encapsidated viroid-like RNA, *Plant Pathol.*, 32, 47, 1983.
49. **Paliwal, Y. C.,** Properties of a Canadian isolate of lucerne transient streak virus and further evidence of similarity of the virus to sobemo-viruses, *Can. J. Plant Pathol.*, 6, 1, 1984.
50. **Matthews, R. E. F.,** Classification and nomenclature of viruses, *Intervirology*, 17, 1, 1982.
51. **Gould, A. R., Francki, R. I. B., and Randles, J. W.,** Studies on encapsidated viroid-like RNA. IV. Requirement for infectivity and specificity of two RNA components from velvet tobacco mottle virus, *Virology*, 110, 420, 1981.
52. **Jones, A. T., Mayo, M. A., and Duncan, G. H.,** Satellite-like properties of small circular RNA molecules in particles of lucerne transient streak virus, *J. Gen. Virol.*, 64, 1167, 1983.
53. **Jones, A. T. and Mayo, M. A.,** Satellite nature of the viroid-like RNA-2 of solanum nodiflorum mottle virus and the ability of other plant viruses to support the replication of viroid-like RNA molecules, *J. Gen Virol.*, 65, 1713, 1984.
54. **Francki, R. I. B.,** Personal communication.
55. **Bruening, G.,** Personal communication.
56. **Haseloff, J.,** Structure of Viroids, Virusoids and Satellites, Ph.D. thesis, University of Adelaide, Adelaide, Australia, 1983.
57. **Symons, R. H., Haseloff, J., Visvader, J. E., Keese, P., Murphy, P. J., Gill, D. S., Gordon, K. H. J., and Bruening, G.,** On the mechanism of replication of viroids, virusoids, and satellite RNAs, in *Subviral Pathogens of Plants and Animals: Viroids and Prions*, Maramorosch, K. and McKelvey, J. J., Jr., Eds., Academic Press, Orlando, Fla., 1985, 235.
58. **Kiefer, M. C., Daubert, S. D., Schneider, I. R., and Bruening, G.,** Multimeric forms of satellite of tobacco ringspot virus RNA, *Virology*, 121, 262, 1982.
59. **Linthorst, H. J. M. and Kaper, J. M.,** Circular satellite-RNA molecules in satellite of tobacco ringspot virus-infected tissue, *Virology*, 137, 206, 1984.
60. **Sogo, J. M. and Schneider, I. R.,** Electron microscopy of double-stranded nucleic acids found in tissue infected with the satellite of tobacco ringspot virus, *Virology*, 117, 401, 1982.
61. **Hutchins, C. J., Keese, P., Visvader, J. E., Rathjen, P. D., McInnes, J. L., and Symons, R. H.,** Comparison of multimeric plus and minus forms of viroids and virusoids, *Plant Mol. Biol.*, 4, 293, 1985.
62. **Branch, A. D., Robertson, H. D., and Dickson, E.,** Longer-than-unit-length viroid minus strands are present in RNA from infected plants, *Proc. Natl. Acad. Sci. U.S.A.*, 78, 6381, 1981.
63. **Chu, P. W. G., Francki, R. I. B., and Randles, J. W.,** Detection, isolation and characterization of high molecular weight double-stranded RNAs in plants infected with velvet tobacco mottle virus, *Virology*, 126, 480, 1983.
64. **Muhlbach, H.-P., Faustmann, O., and Sänger, H. L.,** Conditions for optimal growth of a PSTV-infected potato cell suspension and detection of viroid-complementary longer-than-unit-length RNA in these cells, *Plant Mol. Biol.*, 2, 239, 1983.
65. **Owens, R. A. and Diener, T. O.,** RNA intermediates in potato spindle tuber viroid replication, *Proc. Natl. Acad. Sci. U.S.A.*, 79, 113, 1982.
66. **Ishikawa, M., Meshi, T., Ohno, T., Okada, Y., Sano, T., Veda, I., and Shikata, E.,** A revised replication cycle for viroids: the role of longer than unit length RNA in viroid replication, *Mol. Gen. Genet.*, 196, 421, 1984.
67. **Imperial, J. S., Rodriguez, M. J. B., and Randles, J. W.,** Variation in the viroid-like RNA associated with cadang-cadang disease: evidence for an increase in molecular weight with disease progress, *J. Gen. Virol.*, 56, 77, 1981.
68. **Mohamed, N. A., Haseloff, J., Imperial, J. S., and Symons, R. H.,** Characterization of the different electrophoretic forms of the cadang-cadang viroid, *J. Gen. Virol.*, 63, 181, 1982.
69. **Bruening, G., Gould, A. R., Murphy, P. J., and Symons, R. H.,** Oligomers of avocado sunblotch viroid are found in infected avocado leaves, *FEBS Lett.*, 148, 71, 1982.
70. **Haseloff, J., Keese, P., and Symons, R. H.,** Unpublished data.
71. **Diener, T. O., Hadidi, A., and Owens, R. A.,** Methods for studying viroids, in *Methods in Virology*, Vol. 6, Maramorosch, K. and Koprowski, H., Eds., Academic Press, New York, 1977, 185.
72. **Semancik, J. S., Morris, T. J., Weathers, L. G., Rodorf, B. F., and Kearns, D. R.,** Physical properties of a minimal infectious RNA (viroid) associated with the exocortis disease, *Virology*, 63, 160, 1975.
73. **Niblett, C. L., Dickson, E., Horst, R. K., and Romaine, C. P.,** Additional hosts and an efficient purification procedure for four viroids, *Phytopathology*, 70, 610, 1980.
74. **Palukaitis, P. and Symons, R. H.,** Purification and characterization of the circular and linear forms of chrysanthemum stunt viroid, *J. Gen. Virol.*, 46, 477, 1980.
75. **Symons, R. H.,** The two-step purification of ribosomal RNA and plant viral RNA by polyacrylamide slab gel electrophoresis, *Aust. J. Biol. Sci.*, 31, 25, 1978.

76. **Kay, E. R. M., Simmons, N. S., and Dounce, A. L.**, An improved preparation of sodium desoxyribonucleate, *J. Am. Chem. Soc.*, 74, 1724, 1952.
77. **Barker, J. M., McInnes, J. L., Murphy, P. J., and Symons, R. J.**, Dot-blot procedure with [^{32}P]DNA probes for the sensitive detection of avocado sunblotch and other viroids in plants, *J. Virol. Methods*, 10, 87, 1985.
78. **Laulhere, J.-P. and Rozier, C.**, One-step extraction of plant nucleic acids, *Plant Sci. Lett.*, 6, 237, 1976.
79. **Colpan, M., Schumacher, J., Brüggemann, W., Sänger, H. L., and Riesner, D.**, Large-scale purification of viroid RNA using Cs_2SO_4 gradient centrifugation and high-performance liquid chromatography, *Anal. Biochem.*, 131, 257, 1983.
80. **Schwinghamer, M.**, Personal communication.
81. **Da Graca, J. V. and Mason, T. E.**, Detection of avocado sunblotch viroid in flower buds by polyacrylamide gel electrophoresis, *Phytopathol. Z.*, 108, 262, 1983.
82. **Imperial, J. S. and Rodriguez, M. J. B.**, Personal communication.
83. **Symons, R. H., Barker, J. M., and Visvader, J. E.**, Unpublished data.
84. **Visvader, J. E. and Symons, R. H.**, Unpublished data.
85. **Owens, R. A., Erbe, E., Hadidi, A., Steere, R. L., and Diener, T. O.**, Separation and infectivity of circular and linear forms of potato spindle tuber viroid, *Proc. Natl. Acad. Sci. U.S.A.*, 74, 3859, 1977.
86. **Maniatis, T., Fritsch, E. F., and Sambrook, J.**, *Molecular Cloning. A Laboratory Manual*, Cold Spring Harbor Laboratory, Cold Spring Harbor, N.Y., 1982.
87. **Melton, D. A., Krieg, P. A., Rebagliati, M. R., Maniatis, T., Zinn, K., and Green, M. R.**, Efficient *in vitro* synthesis of biologically active RNA and RNA hybridization probes from plasmids containing a bacteriophage SP6 promoter, *Nucleic Acids Res.*, 12, 7035, 1984.
88. **Tabor, S. and Richardson, C. C.**, A bacteriophage T7 RNA polymerase/promoter system for controlled exclusive expression of specific genes, *Proc. Natl. Acad. Sci. U.S.A.*, 82, 1074, 1985.
89. **Hutchins, C. J., Rathjen, P. D., Forster, A. C., and Symons, R. H.**, Self-cleavage of plus and minus RNA transcripts of avocado sunblotch viroid, *Nucl. Acids Res.*, 14, 3267, 1986.
90. **Sänger, H. L., Klotz, G., Riesner, D., Gross, H. J., and Kleinschmidt, A. K.**, Viroids are single-stranded covalently closed circular RNA molecules existing as highly base-paired rod-like structures, *Proc. Natl. Acad. Sci. U.S.A.*, 73, 3852, 1976.
91. **McClements, W. L. and Kaesberg, P.**, Size and secondary structure of potato spindle tuber viroid, *Virology*, 76, 477, 1977.
92. **Randles, J. W. and Hatta, T.**, Circularity of the ribonucleic acids associated with cadang-cadang disease, *Virology*, 96, 47, 1979.
93. **Ohno, T., Akiya, J., Higuchi, M., Okada, Y., Yoshikawa, N., Takahashi, T., and Hashimoto, J.**, Purification and characterization of hop stunt viroid, *Virology*, 118, 54, 1982.
94. **Tinoco, I., Jr., Uhlenbeck, O. C., and Levine, M. D.**, Estimation of secondary structure in ribonucleic acids, *Nature (London)*, 230, 362, 1971.
95. **Riesner, D., Colpan, M., Goodman, T. C., Nagel, L., Schumacher, J., Steger, G., and Hofmann, H.**, Dynamics and interactions of viroids, *J. Biomol. Struct. Dyn.*, 1, 669, 1983.
96. **Steger, G., Hofmann, H., Förtsch, J., Gross, H. J., Randles, J. W., Sänger, H. L., and Riesner, D.**, Conformational transitions in viroids and virusoids: comparison of results from energy minimization algorithm and from experimental data, *J. Biomol. Struct. Dyn.*, 2, 543, 1984.
97. **Nussinov, R. and Jacobson, A. B.**, Fast algorithm for predicting the secondary structure of single-stranded RNA, *Proc. Natl. Acad. Sci. U.S.A.*, 77, 6309, 1980.
98. **Zuker, M. and Stiegler, P.**, Optimal computer folding of large RNA sequences using thermodynamics and auxiliary information, *Nucleic Acids Res.*, 9, 133, 1981.
99. **Comay, E., Nussinov, R., and Comay, O.**, An accelerated algorithm for calculating the secondary structure of single stranded RNAs, *Nucleic Acids Res.*, 12, 53, 1984.
100. **Henco, K., Steger, G., and Riesner, D.**, Melting curves on less than 1 μg of nucleic acid, *Anal. Biochem.*, 101, 225, 1980.
101. **Henco, K., Sänger, H. L., and Riesner, D.**, Fine structure melting of viroids as studied by kinetic methods, *Nucleic Acids Res.*, 6, 3041, 1979.
102. **Randles, J. W., Steger, G., and Riesner, D.**, Structural transitions in viroid-like RNAs associated with cadang-cadang disease, velvet tobacco mottle virus, and *Solanum nodiflorum* mottle virus, *Nucleic Acids Res.*, 10, 5569, 1982.
103. **Riesner, D. and Römer, R.**, Thermodynamics and kinetics of conformational transitions in oligonucleotides and tRNA, in *Physicochemical Properties of Nucleic Acids*, Vol. 2, Duchesne, J., Ed., Academic Press, London, 1973, 237.
104. **Langowski, J., Henco, K., Riesner, D., and Sänger, H. L.**, Common structural features of different viroids: serial arrangement of double helical sections and internal loops, *Nucleic Acids Res.*, 5, 1589, 1978.

105. **Keese, P. and Symons, R. H.,** Domains in viroids: evidence of intermolecular RNA rearrangements and their contribution to viroid evolution, *Proc. Natl. Acad. Sci. U.S.A.*, 82, 4582, 1985.
106. **Kallenbach, N. R.,** Theory of thermal transitions in low molecular weight RNA chains, *J. Mol. Biol.*, 37, 445, 1968.
107. **Richards, K. E., Jonard, G., Jacquemond, M., and Lot, H.,** Nucleotide sequence of cucumber mosiac virus-associated RNA 5, *Virology*, 89, 395, 1978.
108. **Riesner, D., Kaper, J. M., and Randles, J. W.,** Stiffness of viroids and viroid-like RNA in solution, *Nucleic Acids Res.*, 10, 5587, 1982.
109. **Saenger, W.,** *Principles of Nucleic Acid Structure*, Springer-Verlag, New York, 1983.
110. **Branch, A. D., Benenfeld, B. J., and Robertson, H. D.,** Unusual properties of two branched RNAs with circular and linear components, *Nucleic Acids Res.*, 13, 4889, 1985.
111. **Erdmann, V. A., Wolters, J., Huysmans, E., Vandenberghe, A., and De Wachter, R.,** Collection of published 5S and 5.8S ribosomal RNA sequences, *Nucleic Acids Res.*, 12, 133, 1984.
112. **Visvader, J. E., Forster, A. C., and Symons, R. H.,** Infectivity and *in vitro* mutagenesis of monomeric cDNA clones of citrus exocortis viroid indicates the site of processing of viroid precursors, *Nucleic Acids Res.*, 13, 5843, 1985.
113. **Mikelsaar, R.,** C-A base pairs in transfer ribonucleic acids and codon-anticodon complexes, *J. Theor. Biol.*, 92, 163, 1981.
114. **Dickson, E., Prensky, W., and Robertson, H. D.,** Comparative studies of two viroids: analysis of potato spindle tuber and citrus exocortis viroids by RNA fingerprinting and polyacrylamide-gel electrophoresis, *Virology*, 68, 309, 1975.
115. **Gross, H. J., Domdey, H., and Sänger, H. L.,** Comparative oligonucleotide fingerprints of three plant viroids, *Nucleic Acids Res.*, 4, 2021, 1977.
116. **Dickson, E., Diener, T. O., and Robertson, H. D.,** Potato spindle tuber and citrus exocortis viroids undergo no major sequence changes during replication in two different hosts, *Proc. Natl. Acad. Sci. U.S.A.*, 75, 951, 1978.
117. **Dickson, E., Robertson, H. D., Niblett, C. L., Horst, R. K., and Zaitlin, M.,** Minor differences between nucleotide sequences of mild and severe strains of potato spindle tuber viroid, *Nature (London)*, 277, 60, 1979.
118. **Gross, H. J., Liebl, U., Alberty, H., Krupp, G., Domdey, H., Ramm, K., and Sänger, H. L.,** A severe and a mild potato spindle tuber viroid isolate differ in three nucleotide exchanges only, *Biosci. Rep.*, 1, 235, 1981.
119. **Frisby, D.,** Oligonucleotide mapping of non-radioactive virus and messenger RNAs, *Nucleic Acids Res.*, 4, 2975, 1977.
120. **De Wachter, R. and Fiers, W.,** Preparative two-dimensional polyacrylamide gel electrophoresis of ^{32}P-labelled RNA, *Anal. Biochem.*, 49, 184, 1972.
121. **England, T. E. and Uhlenbeck, O. C.,** 3′-Terminal labelling of RNA with T4 RNA ligase, *Nature (London)*, 275, 560, 1978.
122. **Donis-Keller, H., Maxam, A. M., and Gilbert, W.,** Mapping adenines, guanines and pyrimidines in RNA, *Nucleic Acids Res.*, 4, 2527, 1977.
123. **Peattie, D. A.,** Direct chemical method for sequencing RNA, *Proc. Natl. Acad. Sci. U.S.A.*, 76, 1760, 1979.
124. **Simoncsits, A., Brownlee, G. G., Brown, R. S., Rubin, J. R., and Guilley, H.,** New rapid gel sequencing method for RNA, *Nature (London)*, 269, 833, 1977.
125. **Pilly, D., Niemeyer, A., Schmidt, M., and Bargetzi, J. P.,** Enzymes for RNA sequence analysis, preparation and specificity of exoplasmodial ribonucleases I and II from *Physarym polycephalum*, *J. Biol. Chem.*, 253, 437, 1978.
126. **Donis-Keller, H.,** Phy M: an RNase activity specific for U and A nucleotides useful in RNA sequence analysis, *Nucleic Acids Res.*, 8, 3133, 1980.
127. **Lockard, R. E., Alzner-Deweerd, B., Heckman, J. E., MacGee, J., Tabor, M. W., and RajBhandary, U. L.,** Sequence analysis of 5′[^{32}P]-labelled mRNA and tRNA using polyacrylamide gel electrophoresis, *Nucleic Acids Res.*, 5, 37, 1978.
128. **Boguski, M. S., Hieter, P. A., and Levy, C. C.,** Identification of a cytidine-specific ribonuclease from chicken liver, *J. Biol. Chem.*, 255, 2160, 1980.
129. **Kramer, F. R. and Mills, D. R.,** RNA sequencing with radioactive chain-terminating ribonucleotides, *Proc. Natl. Acad. Sci. U.S.A.*, 75, 5334, 1978.
130. **Maxam, A. M. and Gilbert, W.,** A new method for sequencing DNA, *Proc. Natl. Acad. Sci. U.S.A.*, 74, 560, 1977.
131. **Zimmern, D. and Kaesberg, P.,** 3′-Terminal nucleotide sequence of encephalomyocarditis virus RNA determined by reverse transcriptase and chain-terminating inhibitors, *Proc. Natl. Acad. Sci. U.S.A.*, 75, 4257, 1978.

132. **Symons, R. H.,** Extensive sequence homology at the 3′-termini of the four RNAs of cucumber mosaic virus, *Nucleic Acids Res.*, 7, 825, 1979.
133. **Owens, R. A. and Cress, D. E.,** Molecular cloning and characterization of potato spindle tuber viroid cDNA sequences, *Proc. Natl. Acad. Sci. U.S.A.*, 77, 5302, 1980.
134. **van Wezenbeek, P., Vos, P., van Boom, J., and van Kammen, A.,** Molecular cloning and characterization of a complete DNA copy of potato spindle tuber viroid RNA, *Nucleic Acids Res.*, 10, 7947, 1982.
135. **Cress, D. E., Kiefer, M. C., and Owens, R. A.,** Construction of infectious potato spindle tuber viroid cDNA clones, *Nucleic Acids Res.*, 11, 6821, 1983.
136. **Tabler, M. and Sänger, H. L.,** Cloned single- and double-stranded DNA copies of potato spindle tuber viroid (PSTV) RNA and co-inoculated subgenomic DNA fragments are infectious, *EMBO J.*, 3, 3055, 1984.
137. **Sanger, F., Coulson, A. R., Barrell, B. G., Smith, A. J. H., and Roe, B. A.,** Cloning in single-stranded bacteriophage as an aid to rapid DNA sequencing, *J. Mol. Biol.*, 143, 161, 1980.
138. **Okayama, H. and Berg, P.,** High-efficiency cloning of full-length cDNA, *Mol. Cell. Biol.*, 2, 161, 1982.
139. **Meshi, T., Ishikawa, M., Watanabe, Y., Yamaya, J., Okada, Y., Sano, T., and Shikata, E.,** The sequence necessary for the infectivity of hop stunt viroid cDNA clones, *Mol. Gen. Genet.*, 200, 199, 1985.
140. **Peattie, D. A., Douthwaite, S., Garrett, R. A., and Noller, H. F.,** A "bulged" double helix in a RNA-protein contact site, *Proc. Natl. Acad. Sci. U.S.A.*, 78, 7331, 1981.
141. **Christiansen, J., Douthwaite, S. R., Christensen, A., and Garrett, R. A.,** Does unpaired adenosine-66 from helix II of *Escherichia coli* 5S RNA bind to protein L18?, *EMBO J.*, 4, 1019, 1985.
142. **Runia, W. Th. and Peters, D.,** The response of plant species used in agriculture and horticulture to viroid infections, *Neth. J. Plant Pathol.*, 86, 135, 1980.
143. **Imperial, J. S., Bautista, R. M., and Randles, J. W.,** Transmission of the coconut cadang-cadang viroid to six species of plam by inoculation with nucleic acid extracts, *Plant Pathol.*, in press, 1985.
144. **Kikuchi, Y., Tyc, K., Filipowicz, W., Sänger, H. L., and Gross, H. J.,** Circularization of linear viroid RNA via 2′-phosphomonoester, 3′, 5′-phosphodiester bonds by a novel type of RNA ligase from wheat germ and *Chlamydomonas*, *Nucleic Acids Res.*, 10, 7521, 1982.
145. **Palukaitis, P. and Zaitlin, M.,** The nature and biological significance of linear potato spindle tuber viroid molecules, in *Plant Infectious Agents. Viruses, Viroids, Virusoids and Satellites*, Robertson, H. D., Howell, H. S., Zaitlin, M., and Malmberg, R. K., Eds., Cold Spring Harbor Laboratory, Cold Spring Harbor, N.Y., 1983, 136.
146. **Goodman, T. C., Nagel, L., Rappold, W., Klotz, G., and Riesner, D.,** Viroid replication: equilibrium association constant and comparative activity measurements for the viroid-polymerase interaction, *Nucleic Acids Res.*, 12, 6231, 1984.
147. **Schumacher, J., Sänger, H. L., and Riesner, D.,** Subcellular localization of viroids in highly purified nuclei from tomato leaf tissue, *EMBO J.*, 2, 1549, 1983.
148. **Haseloff, J.,** Personal communication.
149. **Kiberstis, P. A., Haseloff, J., and Zimmern, D.,** 2′-Phosphomonoester, 3′-5′ phosphodiester bond at a unique site in a circular viral RNA, *EMBO J.*, 4, 817, 1985.
150. **Konarska, M., Filipowicz, W., Domdey, H., and Gross, H. J.,** Formation of a 2′-phosphomonoester, 3′,5′-phosphodiester linkage by a novel RNA ligase in wheat germ, *Nature (London)*, 293, 112, 1981.
151. **Jones, A. T. and Mayo, M. A.,** Interaction of lucerne transient streak virus and the viroid-like RNA-2 of *Solanum nodiflorum* mottle virus, *J. Gen. Virol.*, 64, 1771, 1983.
152. **Paliwal, Y. C.,** Interaction of the viroid-like RNA-2 of lucerne transient streak virus with southern bean mosiac virus, *Can. J. Plant Pathol.*, 6, 93, 1984.
153. **Francki, R. I. B., Chu, P. W. G., and Keese, P. K.,** The satellite nature of a viroidlike RNA from lucerne transient streak virus, in *Plant Infectious Agents. Viruses, Viroids, Virusoids and Satellites*, Robertson, H. D., Howell, S. H., Zaitlin, M., and Malmberg, R. L., Eds., Cold Spring Harbor Laboratory, Cold Spring Harbor, N.Y., 1983, 175.
154. **Hillar, M., Stolzmann, Z., Wagle, J., Cicconi, F., Marmocchi, G., and Wyborny, L. E.,** Nuclear deprimerones (low molecular weight peptides controlling gene expression) are associated with DNA and nuclear RNA, *Mol. Biol. Rep.*, 8, 157, 1982.
155. **Davies, J. W., Kaesberg, P., and Diener, T. O.,** Potato spindle tuber viroid. XII. An investigation of viroid RNA as a messenger for protein synthesis, *Virology*, 61, 281, 1974.
156. **Hall, T. C., Wepprich, R. K., Davies, J. W., Weathers, L. G., and Semancik, J. S.,** Functional distinctions between the ribonucleic acids from citrus exocortis viroid and plant viruses: cell-free translation and aminoacylation reactions, *Virology*, 61, 486, 1974.
157. **Semancik, J. S., Conejero, V., and Gerhart, J.,** Citrus exocortis viroid: survey of protein synthesis in *Xenopus laevis* oocytes following addition of viroid RNA, *Virology*, 80, 218, 1977.
158. **Kozak, M.,** Inability of circular mRNA to attach to eukaryotic ribosomes, *Nature (London)*, 280, 82, 1979.

159. **Conejero, V. and Semancik, J. S.,** Exocortis viroid: alteration in the proteins of *Gynura aurantiaca* accompanying viroid infection, *Virology,* 77, 221, 1977.
160. **Zelcer, A., van Adelsberg, J., Leonard, D. A., and Zaitlin, M.,** Plant cell suspension cultures sustain long-term replication of potato spindle tuber viroid, *Virology,* 109, 314, 1981.
161. **Sherman, F., McKnight, G., and Stewart, J. W.,** AUG is the only initiation codon in eukaryotes, *Biochim. Biophys. Acta,* 609, 343, 1980.
162. **Prody, G. A., Bakos, J. T., Buzayan, J. M., Schneider, I. R., and Bruening, G.,** Reversible autolytic processing of dimeric plant virus satellite RNA, *Science,* 231, 1577, 1986.
163. **Rathjen, P. D. and Symons, R. H.,** Unpublished data.
164. **Forster, A. C. and Symons, R. H.,** Unpublished data.
165. **Meshi, T., Ishikawa, M., Ohno, T., Okada, Y., Sano, T., Ueda, I., and Shikata, E.,** Double-stranded cDNAs of hop stunt viroid are infectious, *J. Biochem. (Tokyo),* 95, 1521, 1984.
166. **Zoller, M. J. and Smith, M.,** Oligonucleotide-directed mutagenesis using M13-derived vectors: an efficient and general procedure for the production of point mutations in any fragment of DNA, *Nucleic Acids Res.,* 10, 6487, 1982.
167. **Poncz, M., Solowiejczyk, D., Ballantine, M., Schwartz, E., and Surrey, S.,** "Nonrandom" DNA sequence analysis in bacteriophage M13 by the dideoxy chain-termination method, *Proc. Natl. Acad. Sci. U.S.A.,* 79, 4298, 1982.
168. **Holland, J., Spindler, K., Horodyski, F., Grabau, E., Nichol, S., and VandePol, S.,** Rapid evolution of RNA genomes, *Science,* 215, 1577, 1982.
169. **Reanney, D. C.,** A regulatory role for viral RNA in eukaryotes, *J. Theor. Biol.,* 49, 461, 1975.
170. **Roberts, R. J.,** Intervening sequences excised *in vitro, Nature (London),* 274, 530, 1978.
171. **Crick, F.,** Split genes and RNA splicing, *Science,* 204, 264, 1979.
172. **Diener, T. O.,** Are viroids escaped introns?, *Proc. Natl. Acad. Sci. U.S.A.,* 78, 5014, 1981.
173. **Dickson, E.,** A model for the involvement of viroids in RNA splicing, *Virology,* 115, 216, 1981.
174. **Zimmern, D.,** Do viroids and RNA viruses derive from a system that exchanges genetic information between eukaryotic cells?, *Trends Biochem. Sci.,* 7, 205, 1982.
175. **Collmer, C. W., Hadidi, A., and Kaper, J. M.,** Nucleotide sequence of the satellite of peanut stunt virus reveals structural homologies with viroids and certain nuclear and mitochondrial introns, *Proc. Natl. Acad. Sci. U.S.A.,* 82, 3110, 1985.
176. **Kiss, T. and Solymosy, F.,** Sequence homologies beteen a viroid and a small nuclear RNA (snRNA) species of mammalian origin, *FEBS Lett.,* 144, 318, 1982.
177. **Kiss, T., Posfai, J., and Solymosy, F.,** Sequence homology beteen potato spindle tuber viroid and U3B snRNA, *FEBS Lett.,* 163, 217, 1983.
178. **Krol, A. and Ebel, J.-P.,** U1, U2 and U5 small nuclear RNAs are found in plant cells. Complete nucleotide sequence of the U5 RNA family from pea nuclei, *Nucleic Acids Res.,* 11, 8583, 1983.
179. **Tatei, K., Takemura, K., Mayeda, A., Fujiwara, Y., Tanaka, H., Ishihama, A., and Ohshima, Y.,** U1 RNA-protein complex preferentially binds to both 5′ and 3′ splice junction sequences in RNA or single-stranded DNA, *Proc. Natl. Acad. Sci. U.S.A.,* 81, 6281, 1984.
180. **Zaug, A. J., Grabowski, P. J., and Cech, T. R.,** Autocatalytic cyclization of an excised intervening sequence RNA is a cleavage-ligation reaction, *Nature (London),* 301, 578, 1983.
181. **Dinter-Gottlieb, G.,** Viroids and virusoids are related to group I introns, in *The Viroids,* Diener, T. O., Ed., in press, Plenum Press, New York, 1986.
182. **Cech, T. R.,** RNA splicing: three themes with variations, *Cell,* 34, 713, 1983.
183. **Perrault, J.,** Origin and replication of defective interfering particles, *Curr. Top. Microbiol. Immunol.,* 93, 151, 1981.
184. **Sagata, N., Yasunaga, T., Tsuzuku-Kawamura, J., Ohishi, K., Ogawa, Y., and Ikawa, Y.,** Complete nucleotide sequence of the genome of bovine leukemia virus: its evolutionary relationship to other retroviruses, *Proc. Natl. Acad. Sci. U.S.A.,* 82, 677, 1985.
185. **Haseloff, J., Goelet, P., Zimmern, D., Ahlquist, P., Dasgupta, R., and Kaesberg, P.,** Striking similarities in amino acid sequence among nonstructural proteins encoded by RNA viruses that have dissimilar genomic organization, *Proc. Natl. Acad. Sci. U.S.A.,* 81, 4358, 1984.

Chapter 2

THE VIROID REPLICATION PROCESS

Hugh D. Robertson and Andrea D. Branch

TABLE OF CONTENTS

I. INTRODUCTION

One of the most fascinating challenges posed over the years to microbiologists and pathologists has been the study of agents which rely heavily on host cells in order to replicate. Classical virology has revealed, in animal, plant and bacterial systems, a number of intricate and surprising ways in which disease agents can combine their own small information content with many host factors to stimulate efficient synthesis of their progeny. Some of the most striking examples of this process can be found in small virus systems, in which only a few proteins are encoded in the viral genome. In the RNA bacteriophages, for example, only three proteins are encoded, two of them structural; the third provides one of four subunits of a viral RNA replicase which utilizes three host factors normally concerned with protein synthesis.[1] Smaller still are the plant viroids and related pathogenic RNA systems, which replicate efficiently despite the absence of any coding capacity or detectable association with structural proteins.[2] In this case, the investigator must assume that all functions for viroid replication are provided by the host, with the possible sole exception being one promoted by an RNA-catalyzed enzymatic reaction: it is possible that viroids and related small, self-replicating RNAs could directly catalyze some step or steps of their own replication in a manner analogous to that of other RNAs found capable, for example, of catalyzing RNA cleavage events.[3,4] On balance, however, it is likely that the replication of viroids and related RNAs depends on the closest relationship between the pathogen and its host nucleic acid metabolism of any yet discovered. For this reason, continued study of the viroid replication process is also likely to yield critical insights about RNA metabolism in the host. In this chapter, we will consider recent work on the replication of viroids from the point of view of replication complexes and their component strands, a cycle worked out to explain the steps of viroid replication and the host enzymes which may participate in this cycle. Two additional classes of small RNA molecules, which may share replication strategy with viroids will also be considered: (1) circular viroid-like satellites of plant RNA viruses called "virusoids";[5-7] and (2) several linear RNA satellites of plant viruses which appear to form circular RNAs in vivo.[8-11] In Chapter 1, structural parameters of these two classes were introduced. Here, we will review evidence that they also may participate in a preexisting host-dependent pathway, the existence of which is most clearly defined by a consideration of the minimal requirements for the replication of viroids themselves.

II. NUCLEIC ACID INTERMEDIATES FOR VIROID REPLICATION

A. The Mode of Viroid Replication

From the time Diener and soon thereafter Semancik and Weathers showed that viroids are minute RNA molecules,[12-14] a central question in viroid replication concerned the nucleic acid composition of the template strand for synthesis of viroid progeny. Figure 1 diagrams the possible replication modes which were under consideration.[15] Complementary strands ("minus" strands) composed of RNA were first discovered in 1978 by Grill and Semancik[16] during their studies of citrus exocortis viroid (CEV) and now have also been detected in plants infected by many additional viroids and in plants replicating virusoids and several other satellite RNAs.[17,18] Furthermore, the minus strand of potato spindle tuber viroid (PSTV) appears to contain regions complementary to the entire infectious strand[19] (arbitrarily designated the "plus" strand, despite the absence of demonstrable messenger activity[20-22]) and, therefore, could function in replication. Contrary to the earlier reports,[23,24] experiments by several groups of investigators have established that host plants do not contain detectable amounts of viroid-specific DNA, before or after infection.[25-27] Thus, there is now widespread agreement that viroid replication involves direct RNA to RNA copying.

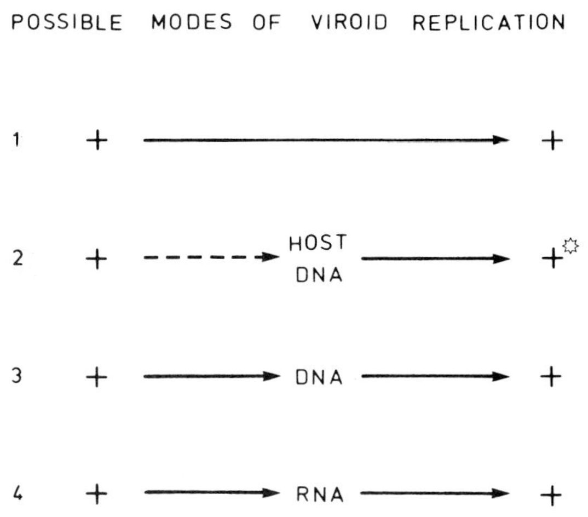

FIGURE 1. Four theoretically possible modes of viroid replication are diagrammed here. In the first mode, no minus strand is involved; replication proceeds directly from one plus strand to another plus strand. No prototypes for this type of replication are known. The autocatalytic mode of replication (mode 2) received a great deal of early attention. It supposes that the infectious plus strand, upon entering a host cell, derepresses an endogenous host gene and causes production of viroid progeny by activating transcription of the host DNA. Since this mode of replication could introduce host-specific changes in the viroid sequence, the progeny plus strand of mode 2 is marked by a star. Alternatively, as in mode 3, the viroid minus strand might be synthesized by reverse transcription of the infectious viroid RNA to create a DNA copy, which would be present only in infected plants. Finally, a replication cycle based on RNA to RNA copying is shown in mode 4. Viroid replication proceeds via an RNA intermediate, which is the template for plus strand synthesis.[15]

One important point established in 1978 was that viroids breed true in different host plants, as found by Dickson et al.,[28] who showed by RNA fingerprinting analysis that PSTV and CEV from tomato and *Gynura* retain their characteristic RNA fingerprint patterns when grown on *Gynura* and tomato, respectively. Thus, replication appears to be dependent on the RNA sequence of the infectious entity, rather than on activation of some latent host DNA or RNA sequence. Subsequent experiments, most notably a series of studies on different viroid species (see Chapter 1) have further strengthened this point of view. Therefore, in seeking to unravel the replication cycle of viroids, the investigator is seeking a means by which an infectious RNA can be copied with fidelity and efficiency, in a manner analogous to more conventional viral replication cycles. The major difference is that no viroid-encoded proteins exist to participate in and direct the reactions. It remains possible that viroids could have rare bases in a pattern imposed by a particular host plant, although no evidence has been found for these as yet. Future work employing direct analysis of viroid RNA labeled in vivo will be required to define such rare bases, if they exist, and their dependence on a particular host. Several observations of "host modification" have been reported.[29] Further study using sequence-pure inocula (rather than naturally occurring viroid isolates) will be needed to reveal the molecular basis of this phenomenon.

B. The Structure of the Viroid Replication Complex

Characterization of the replication complex has played a major role in analyzing replication in a variety of RNA viral systems — ranging from the RNA bacteriophages of *Escherichia coli* to animal and plant viruses. Following deproteinization by phenol, detergents, or other agents, viral plus and minus RNA strands are often found associated in a partially double-, partially single-stranded complex. Early techniques for studying these viral complexes, such as digestion with a variety of ribonucleases and chromatography on cellulose

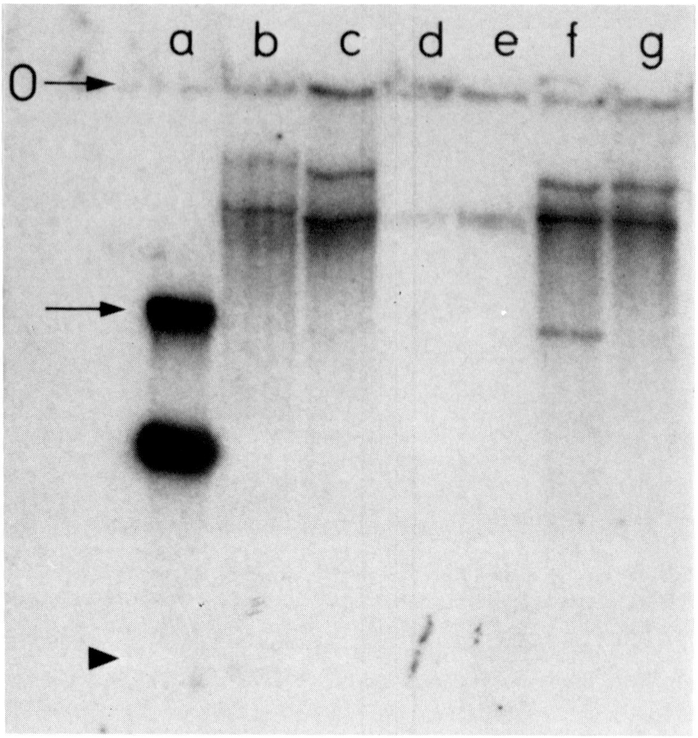

FIGURE 2. Analysis of radioactively labeled PSTV replication complexes. A nucleic acid fraction enriched for double-stranded RNA was prepared from PSTV-infected tomato plants grown in the presence of ^{32}P and ^{33}P as described in the text and in Reference 34. Aliquots were treated in a variety of ways and then fractionated by electrophoresis in a 4% polyacrylamide gel containing 7 *M* urea. The sample in lane (a) contains a mixture of ^{32}P-labeled PSTV circular and linear size markers; those shown in lanes (b to g) contained in vivo labeled nucleic acids after the following treatments: digestion by RNase III[35] (lane b); a control incubation in RNase III buffer[35] (lane c); heat denaturation followed by RNase T1 digestion[30] (lane d); RNase T1 digestion without prior heating (lane e); heat denaturation (lane f); and no pretreatment (lane g); ''O'' marks the origin or electrophoresis; an arrow denotes the position of circular PSTV; and an arrowhead indicates the position of a xylene cyanol blue dye marker.

CF-11 (to separate single- from double- and partially double-stranded RNA) have been combined with Northern hybridization to permit even rare forms of viroid-specific RNAs to be detected in unlabeled populations of plant nucleic acids. In Northern hybridization studies carried out on RNAs fully denatured by chemical modification (as by treatment with glyoxal or formaldehyde) prior to electrophoresis, the structure of the multistranded viroid replication complex was deduced from data about the sizes of single-stranded RNAs released from it. Several studies showed that these complexes are entirely resistant to DNase treatment.[30-33] Studies with ribonucleases on PSTV[30,32] and HSV[33] showed partial resistance of the replication complexes to treatment with enzymes specific for cleavage of single-stranded RNA and, in the case of RNA from PSTV-infected tomato plants, partial sensitivity to a nuclease specific for double-stranded RNA. The replication complex is, therefore, partially single- and partially double-stranded, and most double-stranded regions are about unit length.[30,32]

As illustrated in Figures 2 and 3, it is also possible to use direct RNA analysis to study certain features of the replication complex. For example, double-stranded RNA selected from nucleic acids of plants grown in the presence of radioactive inorganic phosphate[34] can be treated by heat denaturation to release an RNA which co-migrates with the circular form

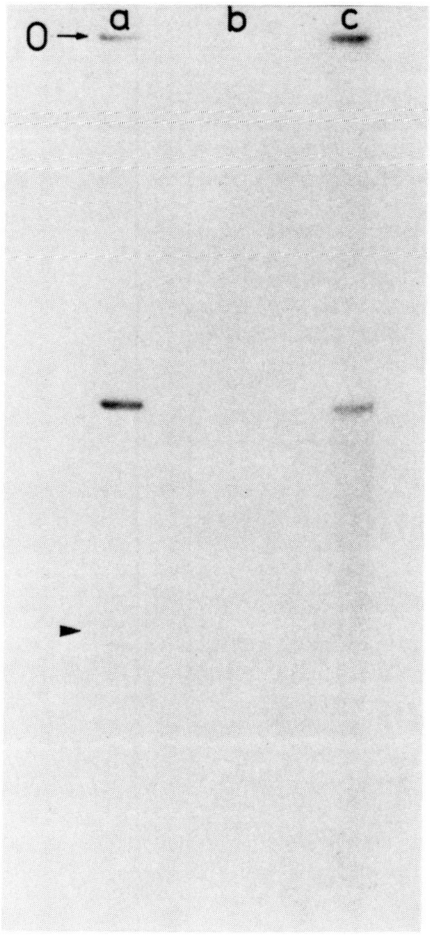

FIGURE 3. Gel electrophoresis of RNAs derived from PSTV replication complexes. The RNase T1-resistant "core" regions of the PSTV replication complex[30] were prepared and labeled in vitro as follows: 2.5 mg of LiCl supernatant fraction of nucleic acids extracted from PSTV-infected tomato plants[2] were digested for 40 min at 37° in 0.01 *M* Tris-HCl, pH 7.6, 0.001 *M* EDTA, and 1 mg/mℓ RNase T1 in a total volume of 0.5 mℓ. At the end of the reaction the salt concentration was raised to 0.1 *M* NaCl and the RNA was then purified by three rounds of phenol extraction and chromatography on cellulose CF-11[31] to enrich for RNA molecules containing double-stranded RNA. Aliquots of RNA prepared in this manner were then either incubated directly with polynucleotide kinase (P-L) and 10 μCi of γ-^{32}P-labeled ATP in a 5 μℓ reaction containing 0.01 *M* Tris-HC1, pH 8.0; 0.01 *M* $MgCl_2$; 0.015 *M* 2-mercaptoethanol for 45 min at 37°; or denatured by heating (in a solution containing 0.01 *M* Tris-HCl, 0.001 *M* EDTA) at 100° for 30 sec, and then incubated with polynucleotide kinase. For a marker, 2.5 ng of PSTV linear RNA were also incubated with polynucleotide kinase in a parallel reaction. 2 μℓ of each sample were then mixed with 2 μg of *E. coli* tRNA carrier and a formamide dye mixture, and then fractionated in 0.3 mm × 40 cm 5% polyacrylamide gels containing 7 *M* urea. Lane (a) contains the linear PSTV size marker. Lanes (b) and (c) contain RNase T1-resistant preparations incubated with polynucleotide kinase in either the native (lane b) or denatured (lane c) state. "O" marks the origin of electrophoresis; the arrowhead denotes the position of a xylene cyanol dye marker.

of PSTV (Figure 2). Thus, while further study is needed to determine the polarity of this RNA, it appears that the PSTV replication complex contains a monomeric, circular RNA hydrogen-bonded to other high molecular weight components as predicted by the rolling circle model of replication shown in Figure 4. The replication complex also contains an RNase-T1 resistant "core" which can be purified by chromatography on cellulose CF-11

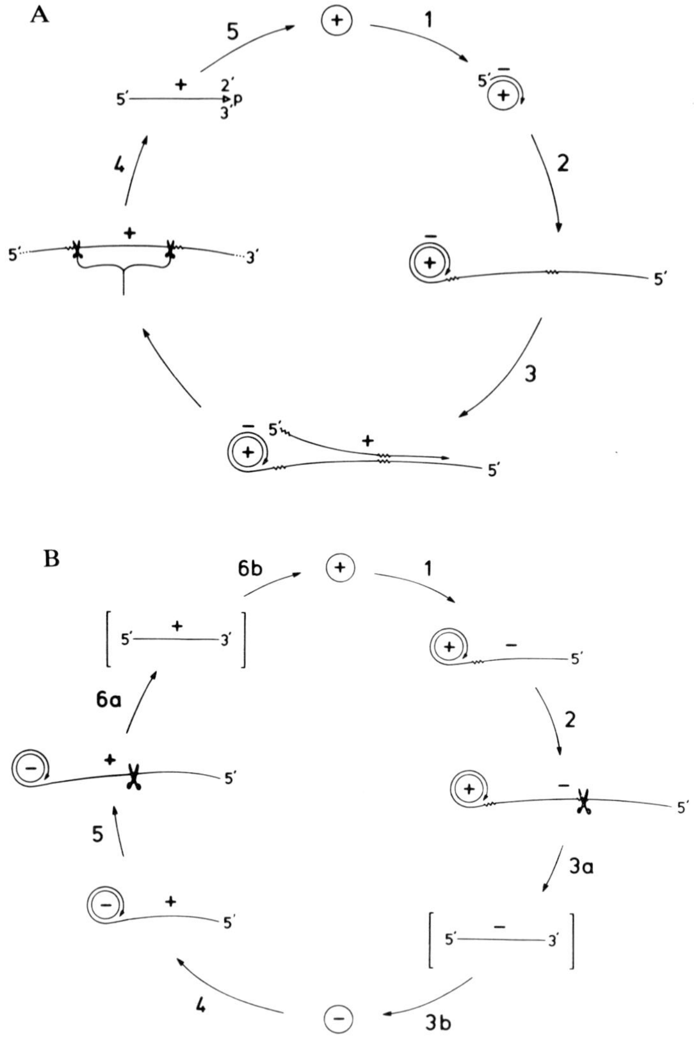

FIGURE 4. Replication cycles for viroid RNA. Two closely related schemes for the replication of circular viroid RNA are presented as described in the text. Both pathways invoke rolling circle synthesis of multimeric RNAs and the cleavage and ligation of multimeric plus strand precursors to yield progeny circles. The cycle in Figure 6A requires a single rolling circle structure, utilizing the input plus strand to synthesize multimeric minus strands; these in turn serve directly as template for multimeric plus strands. The cycle in Figure 6B requires two different rolling circle structures, one as in Figure 6A and the other using circular monomeric minus strands (produced by separate cleavage and ligation events as shown) as template for multimeric plus strand production.

and then radiolabeled in vitro if the two strands in the core are first separated by heat treatment (Figure 3). Both plus and minus strands are labeled by this procedure, but only at a few positions (as shown by dot hybridization and RNA fingerprinting), indicating that the strands may have a rather uniform orientation in the ‘‘core’’.

C. The Rolling Circle Model

In 1981, Branch et al.[30] proposed that viroids use a rolling circle replication scheme in which multimeric minus strands are synthesized by copying of a circular template. One such scheme is shown in Figure 4A, in which the input circular plus strand serves as the template for rolling circle synthesis of multimeric minus strands. These in turn allow synthesis of multimeric plus strands, which are cut down to unit length and circularized in a process reminiscent of eukaryotic RNA splicing. An alternative scheme is shown in Figure 4B, in which the multimeric minus strands indirectly serve as template for plus strands by first undergoing cleavage and ligation to produce circular monomeric minus strands. These in turn serve as template for a second stage of rolling circle synthesis to produce the multimeric plus strands.[17] It is possible that some agents use one variant of this cycle and others use the alternative pathway.

1. Minus Strand Synthesis

Minus strands which are much longer than unit length have been detected in plants replicating a large number of viroids including PSTV,[30-32] CEV,[17,18] ASBV,[18,36] and HSV.[33] In most cases, the minus strands form a prominent series of bands whose lengths are multiples of unit length. There could be several explanations for this observation. First, there could be a pause in RNA synthesis by the polymerase following each round of replication. Alternatively, it is possible that cleavage events take place on the multimeric minus strands. Kinetic studies on the accumulation of minus strand multimers demonstrate that they are detectable as soon as progeny plus strands begin to arise,[17] suggesting that they have an important role in replication. In the case of PSTV and HSV, the lack of detectable levels of monomeric minus strands argues that PSTV and other closely related viroids have a replication cycle like the one shown in Figure 4A; however, information concerning the polarity of the circular monomer released from in vivo-labeled replication complexes (see Figure 2) will help to decide this issue. The observation of monomeric minus strand circles of ASBV[18] indicates that this agent (which has little sequence homology with the other viroids, see Chapter 1) uses the variant cycle diagrammed in Figure 4B.

2. Plus Strand Intermediates

The discovery of multimeric minus strands immediately raised the question of whether plus strands were synthesized directly as unit length species or produced from longer precursors. There is now general agreement that multimeric plus strands are produced and then cleaved during the synthesis of plus strand progeny. Low levels of multimeric plus strands have been detected by Northern hybridization analysis of plants infected by PSTV[15,17,37,38] and CEV,[18] while multimeric forms of CCCV[18] and ASBV[18,36] accumulate to much higher concentrations. When nucleic acids of PSTV-infected plants are enriched for structures containing double-stranded RNA by sucrose gradient sedimentation and cellulose CF-11 chromatography, plus strand dimers and trimers become prominent components, although a significant minority of plus strand monomers is still present,[17] as expected for rolling circle synthesis.

As with the array of multimeric minus strands, it is interesting to speculate why viroid plus strands accumulate principally as integral multiples of unit length. Again, such an observation could be the result of pauses in synthesis after every round of RNA replication, of specific cleavage to remove exact multiples of unit length, or of some combination of the two. The finding of multimeric viroid plus strand precursors is in marked contrast to what is observed in other viruses, e.g., the DNA bacteriophage fl of *E. coli*. In this case, plus strand tails of all possible lengths up to dimer[39] are observed in the replicative intermediate structures. If such were the case with viroids, more RNA strands of fractional length would have been expected leading to a highly pronounced smear of plus strand species between the various integral multiples of unit length. Further studies in vitro or utilizing cultured cells which are replicating viroid RNA[22,40] will be needed to sort out the meaning of this unique aspect of viroid replication.

3. Processing of Multimeric Viroid Transcripts

Both of the models for viroid replication shown in Figure 4 predict an important role for specific RNA cleavage in the life cycle of the viroid. When such specific RNA cleavage occurs reproducibly at the same point or points within an authentic in vivo RNA, it is called RNA processing. Several distinct patterns of RNA processing have been revealed through studies of RNA synthesis in a wide variety of organisms; they provide useful prototypes for the kinds of reactions which might be involved in viroid replication. In comparing viroid replication steps to RNA processing reactions in other systems, it may be helpful to consider the structure of the "natural" linear viroid RNAs which are consistently found in extracts of viroid-infected plants. While the yield of the unit length linear form is highly variable (reports range from barely detectable amounts to levels which are about 15% of the circular form), inoculation studies have proven them to be infectious,[41-43] suggesting that they can be circularized and fed into the replication cycle. In vitro enzymatic studies[44,45] suggest that at least some of the natural linears have 5′ hydroxyl and 2′,3′ cyclic phosphate termini. It is easy to account for these terminal structures if the unit length linear forms are produced by cleavage of a longer precursor; primary transcripts would be expected to have 5′ triphosphate and 3′ hydroxyl moieties. However, since the natural linears could also arise through artifactual nicking of unit length circular forms, the biological role of these linear RNAs is still under investigation.

In vitro study of RNA processing reactions requires relatively large amounts of purified substrates. In the case of ASBV, dimeric forms accumulate and can be purified by gel electrophoresis; initial observations of their autocatalytic cleavage to monomer length RNAs were made using stained gel bands.[36,46] However, in most viroid systems, multimeric plus strands are not present in adequate amounts to be used to study cleavage. Furthermore, detailed analysis of RNA processing is extremely difficult to carry out on unlabeled molecules. For these reasons, multimeric viroid in vitro transcripts were attractive alternatives for biochemical studies; however, before in vitro transcripts could be meaningfully adopted as model RNA processing substrates, it was critical to test their ability to feed into the viroid replication cycle by measuring their infectivity in susceptible host plants.

Several years ago, Owens and colleagues reported the important finding that plasmids containing two copies of viroid cDNA were infectious to plants when inoculated at very high doses and that the resulting viroid RNA had the same sequence as the cDNA, strongly suggesting that natural viroid RNA contains no unusual features (e.g., modified bases) required for replication.[47] Ohno et al.[48] later reported that in vitro RNA transcripts of HSV induced symptoms in plants. In order to study the extent and fidelity of replication, we inoculated plants with dimeric transcripts of PSTV transcribed in vitro with the bacteriophage SP6 polymerase.[34] As illustrated in Figure 5, transcripts contained two head-to-tail copies of PSTV from the strain originally cloned by Van Wezenbeek and co-workers[49] and about 20 bases of flanking sequences from the expression vector. We found that (1) direct SP6 transcripts are infectious to plants — "extra" nonviroid sequences from the vector do not need to be removed prior to inoculation and no specialized structure (such as a 5′ cap) needs to be added; (2) RNA transcripts are approximately as infectious as *bona fide* viroid RNA and have an infectivity several orders of magnitude greater than reported for DNA plasmids containing viroid inserts; (3) RNA transcripts are faithfully copied by host plants, as confirmed by fingerprint analysis of viroid RNA recovered from transcript-inoculated plants. Infectious in vitro viroid transcripts have also been studied by Tabler and Sänger[43] and Visvader et al.[50] From bioassay of HSV, PSTV, and CEV in vitro transcripts, the following additional facts have emerged: (1) unit length transcripts are not infectious, while certain slightly longer than unit length transcripts can induce disease;[43,50] (2) minus strand transcripts are not infectious, or have a much lower infectivity than plus strand transcripts.[33,43]

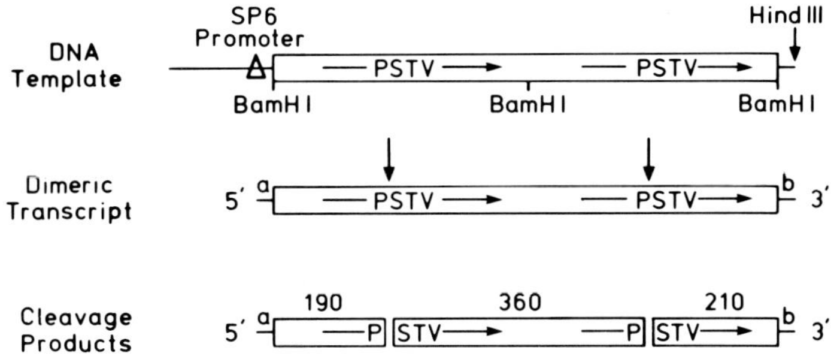

FIGURE 5. Map of the dimeric PSTV cDNA clone and its RNA transcript before and after cleavage. In this scale model, the positions of the bacteriophage SP6 promoter and *Bam*HI restriction sites; the orientation of the PSTV cDNA copies; and the point of cleavage by *Hin*dIII are indicated in the diagram of the DNA template. The boxed regions depict two head-to-tail copies of the PSTV plus strand sequence, marked by horizontal arrows. Regions "a" and "b" designated in the map of the RNA transcript contain vector-specific sequences. Vertical arrows impinging upon the dimeric transcript indicate the regions in which cell-free cleavage occurs. The approximate sizes of the cleavage products are recorded above each segment.[34]

Multimeric transcripts have been tested for self-catalyzed cleavage in a number of laboratories, with differing results.[34,50,51] We incubated ^{32}P-labeled dimeric transcripts of PSTV under a number of conditions and found a low (1 to 5%), but reproducible level of self-catalyzed cleavage. A qualitatively identical pattern of cleavage products was generated when transcripts were incubated in a variety of buffers containing divalent cations. Reaction products included three well-defined bands, one of which co-migrated with unit length PSTV and two smaller fragments. Preliminary mapping of the two smaller fragments suggests that cleavage occurred between positions 250 to 270 of the PSTV map.[52]

However, Visvader et al.[50] found no evidence of a similar reaction in their studies of CEV transcripts, and moreover, cite the infectivity of monomeric transcripts containing a short terminal extension composed of sequences from the central conserved region as an indication that a host endonuclease cleaves in the segment bounded by the bases 89 and 99. Further direct studies are obviously needed to rigorously identify the site of cleavage and to determine the relative importance of autocatalysis and more conventional RNA processing to the cleavage of viroids such as PSTV and CEV. As discussed in detail in Chapter 1, trancripts of ASBV are very efficient substrates for in vitro autocatalysis.

At the moment, it is not at all clear why the transcripts of some agents carry out efficient autocatalytic cleavage in vitro and others only at a low rate. It is possible that mechanistically similar processing reactions take place during the replication of all of these pathogens and that they differ only in the relative effect of host factors on the rate of RNA-catalyzed cleavage. It is estimated that the autocatalytic cleavage of *Tetrahymena* rRNA occurs in vitro at a rate which is about 2% of the in vivo cleavage rate, suggesting that even in this system, other cellular components accelerate the reaction rate. Similar ideas are now being tested in mRNA splicing as well.

One further consideration is relevant to the question of cleavage efficiency and its relation to the generation of infectious viroid progeny RNAs. Symons and his colleagues[53] have noted that ASBV, although its multimeric plus strand forms are capable of efficient self-cleavage in vitro,[18] accumulates high levels of plus strand multimers in vivo; CCCV, on the other hand, accumulates a moderate level of plus strand oligomers; and CEV, which is stated to undergo little or no self-cleavage of plus strand multimers,[50] has the lowest accumulation level of this group of viroids in vivo. Our observations on PSTV, which has very low levels of plus strand multimers in vivo and inefficient self-cleavage of them in vitro,[34,37] go along with this correlation. Symons et al.[53] further relate the accumulation of uncleaved plus strand multimers in vivo to the total amount of viroid present (expressed as a percent of total nucleic acids); again, ASBV is the highest, followed by CCCV and then CEV. Thus, there appears to be a correlation between the dose of viroid building up in the infected cell and the accumulation of uncleaved plus strand oligomers. Symons and coworkers consider the plus multimers to build up because of malfunctioning in the machinery responsible for processing them to monomers. Therefore, the extent of oligomer accumulation may reflect the extent of disruption of the host RNA processing machinery.

These conclusions strongly suggest that even ASBV may require a host factor for efficient processing in vivo, and when this factor is ''titrated,'' there is a buildup of its substrate. This abnormal state may also be related to another set of phenomena observed by Symons and colleagues[18,36,53] — unusual plus and minus strands which are nonintegral multiples of unit length — seen especially in ASBV-infected plants. These fractional multimers, or ''X-forms'', have been suggested by Branch et al.[37] to result from unusual cleavages (at cryptic processing sites), which could occur while the pool of viroid plus strand multimers awaited proper RNA processing. Such incorrect cleavage, taking place beyond the position in the multimeric precursor appropriate to give unit length molecules, could create ''dead end'' products or, if it generated products capable of circularizing and replicating, such cleavage at cryptic sites could lead to the accumulation of molecules containing internal duplications. Multiple rounds of such cryptic cleavage and replication could give rise to species such as those reported by Haseloff et al.[54] for CCCV, in which internal duplications indeed arise. The existence of such forms would seem to underline the importance of an efficient and correct plus strand RNA processing step in the viroid life cycle. In support of this point of view, a recent study by Symons and his colleagues[18] makes a point of relating the RNA multimers of fractional length (the ''X-bands'' mentioned above) to the slow electrophoretic forms of CCCV containing a duplication of one part of the molecule. These authors speculate that the fractional-length multimers are remnants of RNA rearrangements associated with steps of replication or processing.

4. Rolling Circle Replication in Other Small Infectious RNAs

The most striking parallel to the cycles shown in Figure 4 is that proposed for a linear viral satellite RNA, that of tobacco ringspot virus (TobRV), presented by Bruening and colleagues[10] in 1982. Kiefer et al.[10] detected large amounts of multimeric plus and minus strands of TobRV satellite RNA in infected tissues and postulated that they might arise by rolling circle synthesis initiated by circularization of the incoming linear plus strand. Subsequent analyses which support this view are (1) circular forms of the TobRV satellite are indeed present in infected cells;[11] and (2) the infectious linear satellite RNA has 5′ hydroxyl and 2′,3′ cyclic phosphate termini,[55] suggesting that it originates by cleavage from a larger precursor. Kiefer et al.[10] have also found that monomeric minus strands are present in cells infected with TobRV and its satellite, suggesting that these might also play a role in replication. Recent work by Prody and colleagues[55,56] indicates that RNA-catalyzed cleavage may be involved in the production of linear monomers from the multimeric plus strands. In particular, they found that linear monomeric RNAs produced in vitro by autocatalytic cleavage are infectious to plants and that they have the same terminal sequences as naturally occurring linear molecules.

Several lines of evidence suggest that the circular plant satellites (''virusoids''), first discovered in Australia and described by Randles and colleagues,[5-7] have a replication pathway similar to that of viroids, in general, and ASBV, in particular. Longer than unit length virusoid plus and minus strands have been detected in tissue infected by velvet tobacco mottle virus (VTMoV),[57] solanum nodiflorum mottle virus (SNMV)[18] and lucerne transient streak virus (LTSV).[18] Francki et al.[7] report progress toward establishing a cell-free system in which template-independent complexes produce high molecular weight radiolabeled species. However, in comparison to viroids, virusoids replicate to extremely high levels in their host plants. In the case of the virusoid of VTMoV, virusoid RNA accounted for 36 to 50% of the total RNAs present in infected leaves, as reported by Randles et al.[5] Furthermore, virusoid replication has been associated with the cytoplasm,[5,7] while PSTV[38,58,59] and CEV[60,61] replication has been linked to the nucleus.

As noted above for transcripts of ASBV and RNA of TobRV satellite, in vitro transcripts of LTSV virusoid undergo nonenzymatic cleavage.[46] Keese and Symons (Chapter 1) note similarities in the in vitro cleavage sites of these agents. The ability of both plus and minus strand transcripts of ASBV and the LTSV virusoid to promote efficient autocatalytic cleavage under similar conditions in vitro suggests that these agents may have a replication cycle like the one diagrammed in Figure 4B. Kiberstis et al.[62] recently identified a presumptive ligation site in the virusoid of SNMV by mapping the position of a 3′,5′ phosphodiester, 2′ phosphomonoester linkage in mature, circular RNA extracted from virus particles. The presence of the 2′ phosphate suggests that the SNMV virusoid (and probably other closely related RNAs) is cleaved from a precursor in a reaction which generates 5′ hydroxyl and 2′,3′ cyclic phosphate moieties. As noted above, these terminal structures have also been found on cleavage products of ASBV transcripts (plus and minus polarity) and on linear monomers of TobRV-sat[55] and PSTV.[17,44,45]

Finally, an additional class of small viral satellite, those accompanying cucumber mosaic virus (CMV) and peanut stunt virus (PSV), may have replication pathways with many features differing from that of viroids, although they are in the viroid size class (250 to 400 nucleotides). Linthorst and Kaper[63] recently identified multimeric plus strands of the PSV satellite (which is called PARNA-5). While no circular forms of this satellite have been found in nucleic acids of infected plants,[11] it might be fruitful to search for circular molecules in fractions highly enriched for double-stranded RNA. The major form of PARNA-5 and CARNA-5 has a 5′ terminal cap.[64-66] Collmer and Kaper[67] recently showed that a nonencoded G-residue is added during the replication process of CARNA-5.

III. ACTIVITIES REQUIRED FOR VIROID RNA REPLICATION

Studies carried out in a number of laboratories have shown consistently that viroid RNA does not code for any proteins and must depend on the host for any that are needed. Some of the best clues concerning the identity of the host factors involved in viroid replication come from a consideration of the structure of mature viroid RNA and its replication intermediates (see Figure 4). Three types of enzymatic activities are strongly implicated: RNA copying activities, processing enzymes, and ligases. In the case of viroids, it is clear that these activities must reside in the host, or be inherent in the structure of the viroid RNA itself. In the case of satellites, such as the virusoids or the linear satellite of TobRV, the situation is more complex because the small RNA could utilize either host or viral components to catalyze replication steps. Kiefer et al.[10] have speculated, in the case of TobRV satellite RNA, that the properties of viral vs. satellite RNA replication are sufficiently distinct from each other that the satellite may be using a preexisting host pathway (like viroid). Whether this is the case in the systems discussed here will be determined by future experiments, but the similarities between the PSTV replication cycle (Figure 4) and that of the TobRV satellite[10] are very impressive.

A. Enzymes for the Synthesis of Viroid Plus and Minus Strands

The host factors for viroid copying could come from two sources: either macromolecules not normally involved in RNA-dependent RNA synthesis which are recruited for this purpose by the pathogen, or factors that are active in the host cell copying endogenous RNA templates. There is no necessity for the synthesis of plus and minus strands to be carried out by the same enzyme.

Plant cells contain an RNA-dependent RNA polymerase activity which can be assayed in vitro using extracts of both virus-infected and uninfected cells.[68-70] Its in vivo function remains unknown, although Jaspars et al.[71] suggest that it is involved in the initiation of virus RNA synthesis, either acting alone or in concert with viral products. Dorssers et al.[72] have shown that this enzyme is not the one responsible for elongation of viral chains.

In the case of viroids, Boege et al.[73] have shown that an RNA-dependent RNA polymerase from uninfected tomato accepts PSTV RNA as a template, producing some copies which co-migrate with full-length size markers during electrophoresis in denaturing gels. Furthermore, these investigators detected some RNA products in PSTV-stimulated reactions which were greater than unit length. Production of these RNAs was insensitive to α-amanitin concentrations up to 10^{-5} *M*, a significant finding in light of earlier observations reported by Mühlbach and Sänger[74] that viroid RNA synthesis in protoplasts made from CPFV-infected cells appears sensitive to α-amanitin at these concentrations. Furthermore, α-amanitin concentrations of 10^{-8} *M* markedly reduced viroid production in isolated nuclei of CEV-infected cells.[60,61]

On the basis of various drug-inhibition assay and in vitro transcription studies, Sänger[75] has proposed that PSTV minus strand synthesis is carried out by the host cell DNA-dependent RNA polymerase II activity (which normally carries out mRNA synthesis). They found that PSTV is a rather efficient template for the tomato and wheat germ RNA polymerase II in vitro, giving rise to products that co-migrate on denaturing gels with linear PSTV. This reaction is inhibited by 10^{-8} *M* α-amanitin. Data concerning the ionic strength optimum and response to α-amanitin led Semancik and Harper[61] to conclude that RNA polymerase II is the primary enzyme responsible for CEV synthesis in nuclei-rich preparations from infected *Gynura aurantiaca*. These interpretations depend upon a number of assumptions, e.g., that RNA polymerase II has the same drug sensitivities for DNA and RNA templates.

A third activity which has been suggested for involvement with viroid RNA synthesis is the host DNA-dependent RNA polymerase I, normally associated with ribosomal RNA synthesis in the nucleolus. This is an attractive idea because of the finding by Riesner et al.[59] that PSTV RNA accumulates in the nucleolus of infected cells. Furthermore, Palukaitis and Zaitlin[42] have noted homologies between regions of PSTV and RNA polymerase I promoters, but caution is needed in interpreting homologies between DNA and RNA sequences, since their structures can differ significantly. Sequence homologies have also been pointed out between PSTV and the small nuclear RNA, U3, which is found in association with the large ribosomal RNA in the nucleolus.[75,76]

Sänger[75] reports that attempts to get plant RNA polymerase I to copy viroids have so far given only a low yield of small transcription products. Semancik and Harper[61] have reported low levels of α-amanitin resistant CEV RNA synthesis in isolated nuclei of infected cells which could represent the action of RNA polymerase I; Spiesmacher et al.[77] conclude that RNA polymerase I is responsible for copying PSTV minus strands in a preparation of highly purified potato cell nuclei.

Drawing no conclusions about the enzymes involved, Ishikawa et al.[33] propose that minus strand copying is initiated in a region of RNA:RNA duplex structure, because of the enhanced infectivity of multimeric minus strand transcripts (synthesized from an expression vector) complexed to fragments of noninfectious plus strands. Tabler and Sänger[43] found similar results when studying multimeric minus transcripts, but also reported that the addition of a 5′ terminal cap increased infectivity, perhaps through a protective mechanism. Future experiments in this area will undoubtedly make additional use of the new technology for synthesizing viroid-specific transcripts in vitro from expression vectors. In vitro transcripts provide a potential source of purified minus strand template, which could be incubated with a variety of host enzymes, although none of these results in isolation would establish the identity of the enzymes replicating viroid RNA in vivo.

The importance to viroid replication of a macromolecular complex involving DNA as a structural element is currently under investigation. Spiesmacher et al.[77] report that PSTV synthesis is inhibited by DNase treatment of isolated nuclei, while Flores and Semancik[60] find that CEV synthesis is unaltered. More definitive evidence will certainly be needed on this point.

Another way to discern the possible involvement of various host RNA-synthesizing enzymes in viroid replication is to consider the structure of the RNA itself. Ever since the pioneering structural studies carried out by Riesner and colleagues[78] and sequence analysis reported by Gross et al.,[52] most accounts of viroid primary and secondary structure have described it as a collapsed rod with a high proportion of Watson:Crick base pairing (see Chapter 1). It has been argued, in fact, that this configuration may resemble double-stranded DNA, thus attracting host enzymes which normally copy DNA templates.[75] However, recent studies in our laboratory[79] revealed that PSTV contains a structural element which cannot be accounted for by conventional complementary base pairing. Using UV light-induced crosslinking, we found that the central conserved region of viroid has an element of local tertiary structure in which two opposing stretches of unpaired bases (residues 98 to 101 and 256 to 261 in the numbering system of Gross et al.[52]) must actually be wound into a tight structure so that the G at position 98 of PSTV is in extraordinary close apposition to the U at position 260. In the course of comparing this highly characteristic structural element of PSTV with similar elements which preexist in other RNAs, we have been struck by the near identity of this region of PSTV to a region in eukaryotic 5S rRNA which also undergoes a characteristic UV-induced crosslinking between nominally "unpaired" G and U residues.[79] Furthermore, a total of eight bases in the vicinity of the UV crosslink are identical in PSTV and most eukaryotic 5S rRNAs. Several studies[80,81] have suggested that this region of 5S rRNA (between stems IV and V) is involved in the binding of the host transcription factor TFIIIA, whose tight binding to 5S rRNA is responsible for the regulation of that RNAs transcription.[82,83] Normally, the host DNA-dependent RNA polymerase III synthesizes 5S rRNA, but only in conjunction with TFIIIA which must also bind to the DNA template.[84] Since this factor has a higher affinity for the 5S rRNA product,[85] the TFIIIA supply is soon titrated and 5S rRNA synthesis shuts off. Keese and Symons[86] have also suggested that part of the central conserved region may form a protein-binding site, pointing out sequence homologies between the "bulged helix" in this region and the binding site for L 18 protein in *E. coli* 5S rRNA.

Given the remarkable similarity between the primary, secondary, and local tertiary structure of the TFIIIA binding site of 5S rRNA and the central conserved region of the viroid,[79] it is tempting to suggest that PSTV and the other viroids which preserve this region may be engaged in mimicry, with the objective of attracting host TFIIIA. At the moment, it is not known whether 5S rRNA/TFIIIA complexes are ever bound also to RNA polymerase III, but we know that the factor and the enzyme are bound together on the analogous DNA template.[84] These considerations suggest that experiments to test the binding of PSTV plus strands to factor TFIIIA, the ability of PSTV to influence the in vitro synthesis of 5S rRNA from DNA templates, and the effects of TFIIIA on PSTV RNA synthesis by RNA polymerase III would be worthwhile. In this regard, Sänger[75] reports that RNA polymerase III is capable of some initiation on PSTV RNA and can sometimes transcribe the viroid RNA into complete copies. Finally, it should be pointed out that the RNA polymerase III reaction, at least on DNA templates, is resistant to α-amanitin at levels much higher than those needed to inhibit RNA polymerase II.[84] Perhaps minus strands are synthesized by RNA polymerase III, and plus strands by another enzyme. At present, some case can be made for involvement of RNA polymerases I, II, and III and the RNA-dependent activity in viroid replication. The solution of this puzzle, and the identities of the two activities responsible for plus and minus strand synthesis, are awaited with great interest.

Before leaving the subject of RNA synthesis in vitro, we should point out that most RNA viruses whose replication has been studied this way first had crude extracts of their infected cells screened for RNA:protein complexes in which template-independent, viral-specific RNA synthesis was taking place. Usually, such complexes can be purified by gentle techniques such as sucrose density gradient centrifugation and certain kinds of chromatography in nondenaturing solvents. In the case of viroid, either infected plant tissues or permanently infected cell lines might be used for such experiments. Proteins associated with complexes capable of synthesizing viroid-specific RNA could be fractionated and compared to known subunits of all major RNA polymerases of the cell (along with suitable protein standards). The next step might be to attempt subfractionation in a nondestructive manner, to see if template-dependence might be established with plus or minus strand monomers or multimers. Thus, like in vitro protein synthesis, the separate steps of PSTV RNA synthesis could be gradually identified using a complex already carrying out some part of the reaction. Francki and colleagues[7] are already actively applying this approach to their studies of virusoid replication.

B. Enzymatic Activities for the Specific Cleavage and Ligation of Viroid RNA

One way for the last two steps of the viroid replication cycle to occur— cleavage to monomers and ligation to yield circles — would be through reactions analogous to the steps of RNA splicing recently characterized in a variety of eukaryotic systems. In these pathways, segments of an RNA precursor or primary transcript are removed by specific cleavage and the appropriate pieces are joined together to form the mature, functional RNA molecule. In viroid replication, multiple cleavages would be needed to form the linear monomers, while RNA ligation would be needed to form the circles.

There are three RNA splicing pathways which have been studied in some detail, and it is necessary to consider them in this section because it is likely that the viroid may use steps from one or more of them (see Figure 6). Transfer RNA genes from a number of eukaryotes have extra or intervening sequences which are transcribed into RNA and then removed by endonucleolytic cleavage. The pathway for this process has been worked out in yeast (Figure 6A), and appears to be a two-step enzymatic one in which cleavage by a ''splicing ribonuclease'' is followed by ligation by an RNA ligase. A series of studies[87,88] by Abelson and colleagues has led to the characterization of these yeast enzymes. A striking feature of this pathway is the formation by the endonuclease of 5′ hydroxyl and 2′,3′ cyclic phosphate termini. Such termini of the tRNA segments (''exons'') are joined by the RNA ligase in a reaction (first worked out in wheat germ extracts by Konarska and colleagues[89,90]) in which a 2′ phosphate group is created at the junction, in addition to the conventional 3′,5′ phosphodiester bond. In the case of tRNA, additional phosphatase activities are postulated to remove the extra 2′ phosphate before the mature tRNA becomes fully functional.

In contrast to the splicing of yeast and other tRNA precursors by two separate protein enzymes acting in a stepwise fashion, the splicing of *Tetrahymena* ribosomal RNA splicing represents an opposite extreme (Figure 6B). In *Tetrahymena,* Kruger, Cech and colleagues[3,91] have found that the rRNA precursor is autocatalytically cleaved in a transesterification reaction requiring a guanosine cofactor to yield a 3′ hydroxyl terminus on the 5′ exon and releasing the intervening sequence still attached to the 3′ exon, but with the guanosine added to its 5′ end. In the second phase of this concerted reaction, the hydroxyl group on the 5′ exon attacks the 3′ splice site and the two mature segments are joined while the intervening sequence is released as a linear fragment 413 bases long with a 3′ hydroxyl terminus. This linear intervening sequence may then undergo a spontaneous circularization during which time it loses 15 nucleotides from the 5′ end. None of these steps appears to require any proteinaceous components, since they can be carried out in a cell-free system using RNA

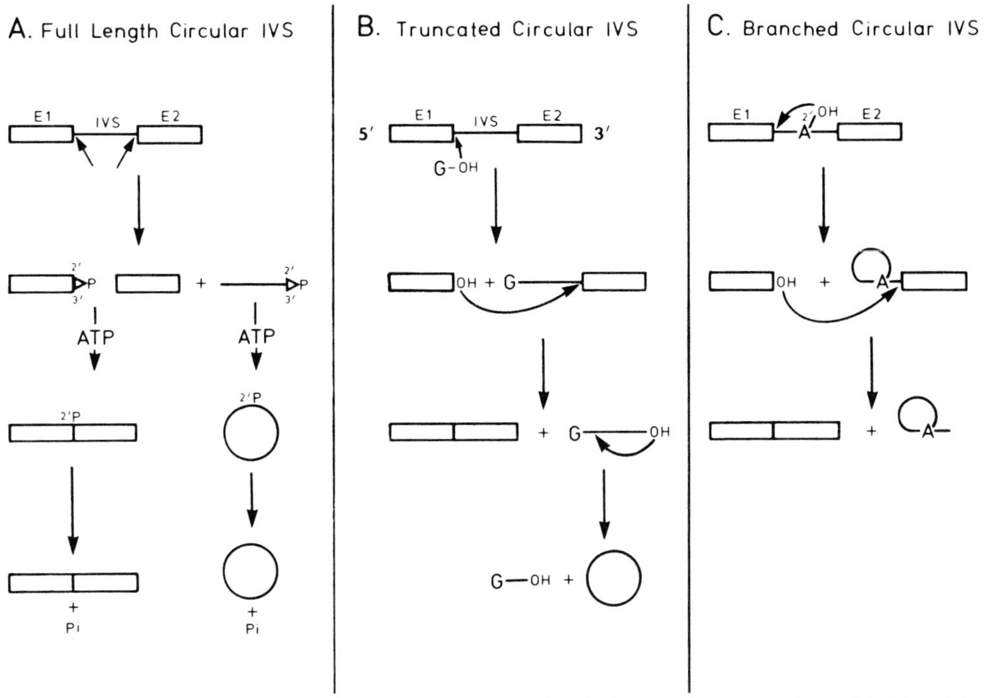

FIGURE 6. Pathways of eukaryotic RNA splicing. As described in the text, three mechanisms of RNA splicing have been discovered. Only the pathway used for splicing tRNA precursors (A) has the potential for preserving the integrity of a genomic sequence if it were represented by the circular element of an excised intervening sequence. The other two mechanisms exclude part of the intervening sequence from the circular portion of either the truncated circular IVS (B) or the branched circular element of the lariat-shaped IVS (C). Replication of either the truncated circle or the circular element of the lariat would result in loss of genetic information.

synthesized in vitro from templates cloned in *E. coli*. As mentioned in a previous section, however, in the cell one or more of these reactions might use protein cofactors to accelerate their rate or broaden their optimal temperature range.[3]

A third cleavage and ligation pathway which has been intensively studied of late (Figure 6C) is that involving the splicing of messenger RNA molecules primarily in mammalian nuclei.[42,43] In this case, a nonlinear form of the intervening sequence arises as a "lariat", in which the 5′ phosphate on the intervening sequence (created by an initial cleavage at the 5′ splice site) becomes attached to the 2′ hydroxyl of a particular A residue 20 to 50 bases upstream from the 3′ splice junction. A second cleavage releases the intervening sequence, which is still in lariat form, and the 3′ exon is then free to join its 5′ terminal counterpart by a mechanism not yet understood. In mammalian systems, best worked out using adenovirus and globin mRNA precursors, as well as in yeast cells, these events are thought to occur on 40 to 60S particles which contain several small nuclear ribonucleoproteins, RNA precursors, and other factors.[94,95] However, recent findings of Peebles and colleagues[96] as well as van der Veen et al.[97] suggest that a yeast mitochondrial mRNA transcript may be able to undergo a similar splicing reaction autocatalytically.

In the case of the specific cleavage and ligation of PSTV RNA, the pathway for splicing of tRNA precursors most resembles that of the viroid, since the excised intervening sequence can be circularized in a reaction which leaves it intact (by the same RNA ligase which joins exon segments to give the mature-sequence tRNA).[88] Both the mRNA and rRNA pathways discussed above lead to nonlinear structures formed from intervening sequences in which part of the intervening sequence is excluded from the circular element. Ever since Roberts[98] first proposed in 1978 that viroids might be "escaped introns", there has been growing evidence for circular structures which arise from intervening sequences during processing. However, in the case of viroids, virusoids, and TobRV satellites, it looks very much as if a linear monomer with 5′ hydroxyl and 2′,3′ cyclic phosphate termini is an intermediate in the synthesis of plus strand progeny. The "splicing ribonuclease" used by yeast to cleave intervening sequences from tRNA precursors creates segments with these terminal structures,[87] and the RNA ligase from wheat germ can then circularize them. It can also circularize natural linear viroid RNAs.[44,45] Interestingly, the RNA ligase from yeast was incapable of circularizing viroid natural linear RNA molecules,[44] despite its ability to complete the second step of yeast tRNA splicing. Thus, there may be as yet undiscovered specificity for enzymatic circularization of linear RNAs like viroid. Since it seems that this step in the viroid life cycle may be catalyzed by a host enzymatic protein, it will probably be necessary to identify and purify the RNA ligase from cells known to be viroid hosts in order to study this reaction under optimal conditions. At the same time, although no evidence of in vitro self-catalyzed circularization of PSTV, ASBV, or any of the satellite RNAs considered here has yet been reported, it is possible that some such reactions may occur. As mentioned above, recent findings in the ASBV, PSTV, and TobRV satellites systems suggest that autocatalytic cleavage can produce 5′ hydroxyl and 2′,3′ cyclic phosphate termini like those depicted in Figure 6A. It would not be surprising if nonenzymatic ligation of such termini could take place, completing an RNA-catalyzed pathway of cleavage and ligation.

Finally, then, what of the cleavage reaction which produces linear viroid monomers? One of the major observations which caused us to study the low efficiency autocatalytic production of PSTV-sized linear plus strands by incubation of dimeric transcripts was the efficient circularization of these molecules by the RNA ligase of wheat germ.[34] The resulting circular RNAs produced by a complete set of in vitro reactions were indistinguishable from natural PSTV infectious circles by a variety of tests including gel electrophoresis on five different systems, and specific enzymatic nicking to confirm their circular configuration. Since we do not know the terminal sequences of PSTV natural linear RNAs found in vivo, we cannot yet compare them directly to those produced by in vitro cleavage. In the case of the TobRV satellite RNA, Prody and colleagues[55,56] have been able to show that their much more efficient cleavage reaction produces the same termini as are found in the mature satellite RNA monomers packaged into virions. More experiments will be required, including incubation of multimeric PSTV transcripts with plant cellular extracts, before any conclusions can be drawn about the relative roles of inherent RNA structure and protein co-factors in the cleavage of PSTV multimeric plus strands. If PSTV is found to require a host protein for optimal cleavage, the "splicing ribonuclease" from the host cells' tRNA splicing pathway should be carefully sought and tested. It seems possible that viroids have adapted a cleavage and ligation pathway also utilized by the host cells' tRNA gene transcripts. Transfer RNAs are among the most highly conserved of RNA molecules. Like 5S rRNA, they are transcribed by RNA polymerase III, and contain regions of local tertiary structure susceptible to UV-induced crosslinking.

IV. DISCUSSION

A general pathway for viroid replication appears to include the synthesis of intermediates (by two host RNA polymerases, one of which may be RNA polymerase III), which are then cleaved and ligated by a pathway reminiscent of tRNA splicing. However, there are many unknowns which indicate the need to identify the host components involved in viroid replication, and then to analyze the role of these host components in uninfected cells. Such investigations may reveal the existence of host analogues to viroid RNA which are copied at the RNA level and also uncover mechanisms of viroid symptom expression. Moreover, studies of viroid pathogenesis must address the question of how viroid replication and pathogenesis are related. Apart from the observation that minor sequence changes in a small portion of the molecule (the "pathogenesis" region, Chapter 1) affect symptoms, the mechanism of viroid pathogenesis remains entirely mysterious.

Many theories have been advanced to explain why sequence changes in the pathogenesis region of PSTV and other viroids lead to altered symptoms.[99] One model assumes that viroid replication and disease induction are separable events. The main experimental support for this model comes from the observation that nucleic acid extracts of plants infected by isolates causing mild, or undetectable, symptoms have about the same steady-state viroid concentration as those infected by severe isolates (usually within the limits of experimental error).[2,75] The pathogenesis region in this model could be an evolutionary remnant, reflecting the origin of the viroid as an escaped regulatory element.[76,100-102] Alternatively, the pathogenesis region could simply be a part of the molecule which is not as tightly constrained by requirements of the replication steps as is most of the rest of the molecule. Thus, naturally arising mutations in this region are permitted to accumulate. An important experimental prediction of this model is that host components exist which interact specifically and differentially with the pathogenesis regions from mild and severe isolates, and that high concentrations of this region would lead to symptoms in the absence of any of the steps of viroid replication.

A second model links replication and pathogenesis tightly together. For example, such a scheme might predict that severe variants reach a concentration which is close to that reached by mild variants, but consistently higher by a margin just sufficient to cause problems for the host. Thus, it might be that severe variants out-compete mild variants for access to the replication machinery. Experimental support for this model comes from our observations that plants develop symptoms of severe disease when inoculated with a mixture in which the mild:severe ratio is 100:1.[99] More efficient replication could be tied to differential symptom expression through two routes: the replication pathway of severe (but not mild) viroid could monopolize an essential host factor, or viroid RNA could be inherently toxic to the host when present at high enough concentrations.

A third model associates the pathogenesis region of PSTV and other viroids with a function which is not essential for replication, but which might confer an advantage to a particular variant under certain growth conditions. The explosive outbreaks of viroid disease leading to agricultural devastation are connected to human intervention (use of contaminated harvesting equipment or clonal propagation of infected stocks). Little is known about the mechanism of viroid spread in the wild, but this third model might predict that the pathogenesis region interacts with a host component important for viroid dispersion under natural conditions, while leaving the efficiency of replication unaltered once it is established. Ultrastructural studies[103-105] show that disease symptoms are accompanied by malformations of the cell wall and associated membranous structures. These changes could be brought about by interaction between viroid RNA and membrane receptors, suggesting that differential binding to transport factors might distinguish mild and severe strains.

Thus, the key question about viroids and similar RNAs in the future will emphasize the cellular factors needed for their replication and the function of their host analogues. Our challenge is to continue seeking such host pathways, if they exist, while sorting out the steps of viroid replication. In this way, it may be possible to find out if and in what way the replication of viroids and their ability to cause disease are related.

ACKNOWLEDGMENTS

The authors would like to thank Drs. G. Bruening, D. L. Rosen, R. H. Symons, and O. C. Uhlenbeck for helpful discussions. We are indebted to the U. S. National Science Foundation for continuing support of our viroid research, currently in the form of NSF grant PCM 83-03943 to H.D.R. and A.D.B.; and for a one-time award for plant growth facilities without which viroid research would be impossible at this location. This research was also supported in part by a McKnight Foundation Individual Award in Plant Molecular Biology to H.D.R. and by USDA grant 85-CRCR-1-1824.

REFERENCES

1. **Zinder, N. D.,** *RNA Phages,* Cold Spring Harbor Laboratory, Cold Spring Harbor, N. Y., 1975.
2. **Diener, T. O.,** *Viroids and Viroid Diseases,* John Wiley & Sons, New York, 1979.
3. **Kruger, K., Grabowski, P. J., Zaug, A. J., Sands, J., Gottschling, D. E., and Cech, T. R.,** Self-splicing RNA: autoexcision and autocyclization of the ribosomal RNA intervening sequence of *Tetrahymena, Cell,* 31, 147, 1982.
4. **Guerrier-Takada, C., Gardiner, K., Marsh, T., Pace, N., and Altman, S.,** The RNA moiety of ribonuclease P is the catalytic subunit of the enzyme, *Cell,* 35, 849, 1983.
5. **Randles, J. W., Davies, C., Hatta, T., Gould, A. R., and Francki, R. I. B.,** Studies on encapsidated viroid-like RNA. I. Characterization of velvet tobacco mottle virus, *Virology,* 108, 111, 1981.
6. **Gould, A. R.,** Studies on encapsidated viroid-like RNA. II. Purification and characterization of a viroid-like RNA associated with velvet tobacco mottle virus (VTMoV), *Virology,* 108, 123, 1981.
7. **Francki, R. I. B., Randles, J. W., Chu, P. W. G., Rohozinski, J., and Hatta, T.,** Viroid-like RNAs incorporated in conventional virus capsids, in *Subviral Pathogens of Plants and Animals: Viroids and Prions,* Maramorosch, K. and McKelvey, J. J., Eds., Academic Press, New York, 1985, 265.
8. **Schneider, I. R.,** Characteristics of a satellite-like virus of tobacco ringspot virus, *Virology,* 45, 108, 1971.
9. **Schneider, I. R. and Thompson, S. M.,** Double-stranded nucleic acids found in tissue infected with the satellite of tobacco ringspot virus, *Virology,* 78, 453, 1977.
10. **Kiefer, M. C., Daubert, S. D., Schneider, I. R., and Bruening, G.,** Multimeric forms of satellite of tobacco ringspot virus RNA, *Virology,* 121, 262, 1982.
11. **Linthorst, H. J. M. and Kaper, J. M.,** Circular satellite RNA molecules in satellite of tobacco ringspot virus-infected tissue, *Virology,* 137, 206, 1984.
12. **Raymer, W. B. and Diener, T. O.,** Potato spindle tuber virus: a plant virus with properties of a free nucleic acid. I. Assay, extraction and concentration, *Virology,* 37, 313, 1969.
13. **Diener, T. O.,** Potato spindle tuber viroid. VIII. Correlation of infectivity with a UV-absorbing component and thermal denaturation properties of the RNA, *Virology,* 50, 606, 1972.
14. **Semancik, J. S. and Weathers, L. G.,** Exocortis virus: an infectious free-nucleic acid plant virus with unusual properties, *Virology,* 47, 456, 1972.
15. **Branch, A. D.,** Characterization of Potato Spindle Tuber Viroid-Specific Nucleic Acids in Tomato Plants, Doctoral dissertation, The Rockefeller University Press, New York, 1981.
16. **Grill, L. K. and Semancik, J. S.,** RNA sequences complementary to citrus exocortis viroid in nucleic acid preparations from infected *Gynura aurantiaca, Proc. Natl. Acad. Sci. U.S.A.,* 75, 896, 1978.
17. **Branch, A. D. and Robertson, H. D.,** A replication cycle for viroids and other small infectious RNAs, *Science,* 223, 450, 1984.
18. **Hutchins, C. J., Keese, P., Visvader, J. E., Rathjen, P. D., McInnes, J. L., and Symons, R. H.,** Comparison of multimeric plus and minus forms of viroids and virusoids, *Plant Mol. Biol.,* 4, 293, 1985.
19. **Zelcer, A., Zaitlin, M., Robertson, H. D., and Dickson, E.,** Potato spindle tuber viroid-infected tissues contain RNA complementary to the entire viroid, *J. Gen. Virol.,* 59, 139, 1982.
20. **Davies, J. W., Kaesberg, P., and Diener, T. O.,** Potato spindle tuber viroid. XII. An investigation of viroid RNA as a messenger for protein synthesis, *Virology,* 61, 281, 1974.
21. **Semancik, J. S., Conejero, V., and Gerhart, J.,** Citrus exocortis viroid: survey of protein synthesis in *Xenopus laevis* oocytes following addition of viroid RNA, *Virology,* 80, 218, 1977.
22. **Zelcer, A., Van Adelsberg, J., Leonard, D. A., and Zaitlin, M.,** Plant cell suspension cultures sustain long-term replication of potato spindle tuber viroid, *Virology,* 109, 314, 1981.
23. **Semancik, J. S. and Geelen, J. L. M. C.,** Detection of DNA complementary to pathogenic viroid RNA in exocortis disease, *Nature (London),* 256, 753, 1975.
24. **Hadidi, A., Jones, D. M., Gillespie, D. H., Wong-Staal, F., and Diener, T. O.,** Hybridization of potato spindle tuber viroid to cellular DNA of normal plants, *Proc. Natl. Acad. Sci. U.S.A.,* 73, 2453, 1976.
25. **Zaitlin, M., Niblett, C. L., Dickson, E., and Goldberg, R. B.,** Tomato DNA contains no detectable regions complementary to potato spindle tuber viroid as assayed by solution and filter hybridization, *Virology,* 104, 1, 1980.

26. **Branch, A. D. and Dickson, E.,** Tomato DNA contains no detectable regions complementary to potato spindle tuber viroid as assayed by Southern hybridization, *Virology,* 104, 10, 1980.
27. **Hadidi, A., Cress, D. E., and Diener, T. O.,** Nuclear DNA from uninfected or potato spindle tuber viroid-infected tomato plants contains no detectable sequences complementary to cloned double-stranded viroid cDNA, *Proc. Natl. Acad. Sci. U.S.A.,* 78, 6932, 1981.
28. **Dickson, E., Diener, T. O., and Robertson, H. D.,** Potato spindle tuber and citrus exocortis viroids undergo no major sequence changes during replication in two different hosts, *Proc. Natl. Acad. Sci. U.S.A.,* 75, 951, 1978.
29. **Peters, D. and Runia, W. T.,** The host range of viroids, in *Subviral Pathogens of Plants and Animals: Viroids and Prions,* Maramorosch, K. and McKelvey, J. J., Eds., Academic Press, New York, 1985, 21.
30. **Branch, A. D., Robertson, H. D., and Dickson, E.,** Longer than unit length viroid minus strands are present in RNA from infected plants, *Proc. Natl. Acad. Sci. U.S.A.,* 78, 6381, 1981.
31. **Rohde, W. and Sänger, H. L.,** Detection of complementary RNA intermediates of viroid replication by Northern blot hybridization, *Biosci. Rep.,* 1, 327, 1981.
32. **Owens, R. A. and Diener, T. O.,** RNA intermediates in potato spindle tuber viroid replication, *Proc. Natl. Acad. Sci. U.S.A.,* 79, 113, 1982.
33. **Ishikawa, M., Meshi, T., Ohno, T., Okada, Y., Sano, T., Ueda, I., and Shikata, E.,** A revised replication cycle for viroids: the role of longer than unit length RNA in viroid replication, *Mol. Gen. Genet.,* 196, 421, 1984.
34. **Robertson, H. D., Rosen, D. L., and Branch, A. D.,** Cell-free synthesis and processing of an infectious dimeric transcript of potato spindle tuber viroid RNA, *Virology,* 142, 441, 1985.
35. **Robertson, H. D., Webster, R. E., and Zinder, N. D.,** Purification and properties of ribonuclease III from *E. coli, J. Biol. Chem.,* 243, 82, 1968.
36. **Bruening, G., Gould, A. R., Murphy, P. J., and Symons, R. H.,** Oligomers of avocado sunblotch viroid are found in infected avocado leaves, *FEBS Lett.,* 148, 71, 1982.
37. **Branch, A. D., Willis, K. K., Davatelis, G., and Robertson, H. D.,** In vivo intermediates and the rolling circle mechanism in viroid replication, in *Subviral Pathogens of Plants and Animals: Viroids and Prions,* Maramorosch, K. and McKelvey, J. J., Eds., Academic Press, New York, 1985, 201.
38. **Spiesmacher, E., Mühlbach, H.-P., Schnölzer, M., Haas, B., and Sänger, H. L.,** Oligomeric forms of potato spindle tuber viroid (PSTV) and of its complementary RNA are present in nuclei isolated from viroid-infected potato cells, *Biosci. Rep.,* 3, 767, 1983.
39. **Zinder, N. D. and Horiuchi, K.,** Multiregulatory element of filamentous bacteriophages, *Microbiol. Rev.,* 49, 101, 1985.
40. **Mühlbach, H.-P., Faustmann, O., and Sänger, H. L.,** Conditions for optimal growth of a PSTV-infected potato cell suspension and detection of viroid-complementary longer-than-unit-length RNA in these cells, *Plant Mol. Biol.,* 2, 239, 1983.
41. **Owens, R. A., Erbe, E., Hadidi, A., Steere, R. L., and Diener, T. O.,** Separation and infectivity of circular and linear forms of potato spindle tuber viroid, *Proc. Natl. Acad. Sci. U.S.A.,* 74, 3859, 1977.
42. **Palukaitis, P. and Zaitlin, M.,** The nature and biological significance of linear potato spindle tuber viroid molecules, in *Plant Infectious Agents,* Robertson, H. D., Howell, S. H., Zaitlin, M., and Malmberg, R. L., Eds., Cold Spring Harbor Laboratory, Cold Spring Harbor, N. Y., 1983, 136.
43. **Tabler, M. and Sänger, H. L.,** Infectivity studies on different potato spindle tuber viroid (PSTV) RNAs synthesized in vitro with the SP6 transcription system, *EMBO J.,* 4, 2191, 1985.
44. **Branch, A. D., Robertson, H. D., Greer, C., Gegenheimer, P., Peebles, C., and Abelson, J.,** Cell-free circularization of viroid progeny RNA by an RNA ligase from wheat germ, *Science,* 217, 1147, 1982.
45. **Kikuchi, Y., Tyc, K., Filipowicz, W., Sänger, H. L., and Gross, H. J.,** Circularization of linear viroid RNA via 2′-phosphomonoester, 3′,5′-phosphodiester bonds by a novel type of RNA ligase from wheat germ and *Chlamydomonas, Nucleic Acids Res.,* 10, 7251, 1982.
46. **Hutchings, C. J., Rathjen, P. J., Forster, A. J., and Symons, R. H.,** Replication of viroids: an in vitro system to characterize the self-processing of oligomeric RNA intermediates of avocado sunblotch viroid to plus and minus monomers, 1st Int. Congr. Plant Mol. Biol., Savannah, Ga., Abstr. OR-13-01, 1985, 30.
47. **Cress, D. E., Kiefer, M. C., and Owens, R. A.,** Construction of infectious potato spindle tuber viroid cDNA clones, *Nucleic Acids Res.,* 11, 6821, 1983.
48. **Ohno, T., Ishikawa, M., Takamatsu, N., Meshi, T., Okada, Y., Sano, S., and Shikata, E.,** In vitro synthesis of infectious RNA molecules from cloned hop stunt viroid complementary DNA, *Proc. Jpn. Acad., Ser. B,* 59, 251, 1983.
49. **Van Wezenbeek, P., Vos, P., van Boom, J., and van Kammen, A.,** Molecular cloning and characterization of a complete DNA copy of potato spindle tuber viroid RNA, *Nucleic Acids Res.,* 10, 7947, 1982.
50. **Visvader, J. E., Forster, A. C., and Symons, R. H.,** Infectivity and in vitro mutagenesis of monomeric cDNA clones of citrus exocortis viroid indicates the site of processing of viroid precursors, *Nucleic Acids Res.,* 13, 5843, 1985.

51. **Tsagris, N., Tabler, M., Mühlbach, H.-P., and Sänger, H. L.,** In vitro synthesized multimeric viroid RNA precursors are processed to monomeric circular viroid RNA in extracts of healthy plant cells, Abstr. Meet. Eukaryotic RNA Processing, Rome, 1985, 65.
52. **Gross, H. J., Domdey, H., Lossow, C., Jank, P., Raba, M., Alberty, H., and Sänger, H. L.,** Nucleotide sequence and secondary structure of potato spindle tuber viroid, *Nature (London),* 273, 203, 1978.
53. **Symons, R. H., Haseloff, J., Visvader, J. E., Keese, P., Murphy, P. J., Gill, D. S., Gordon, K. H. J., and Bruening, G.,** On the mechanism of replication of viroids, virusoids and satellite RNAs, in *Subviral Pathogens of Plants and Animals: Viroids and Prions,* Maramorosch, K. and McKelvey, J. J., Eds., Academic Press, New York, 1985, 235.
54. **Haseloff, J., Mohamed, N. A., and Symons, R. H.,** Viroid RNAs of cadang-cadang disease of coconuts, *Nature (London),* 299, 316, 1982.
55. **Prody, G. A., Bakos, J. T., Buzayan, J. M., Schneider, I. R., and Bruening, G.,** Reversible autolytic processing of dimeric plant virus satellite RNA, *Science,* in press, 1986.
56. **Prody, G. A., Bakos, J. T., Buzayan, J. M., Schneider, I. R., and Bruening, G.,** Self-processing of multimeric forms of satellite tobacco ringspot virus RNA, Abstr. Meet. RNA Processing, Cold Spring Harbor, N. Y., 1984, 8.
57. **Chu, P. W. G., Francki, R. I. B., and Randles, J. W.,** Detection, isolation and characterization of high molecular weight double-stranded RNAs in plants infected with velvet tobacco mottle virus, *Virology,* 126, 480, 1983.
58. **Diener, T. O.,** Potato spindle tuber virus: a plant virus with properties of a free nucleic acid. III. Subcellular location of PSTV-RNA and the question of whether virions exist in extracts or *in situ, Virology,* 43, 75, 1971.
59. **Riesner, D., Colpan, M., Goodman, T. C., Nagel, L., Schumacher, J., and Steger, G.,** Dynamics and interactions of viroids, *J. Biomol. Struct. Dyn.,* 1, 669, 1983.
60. **Flores, R. and Semancik, J. S.,** Properties of a cell-free system for synthesis of citrus exocortis viroid, *Proc. Natl. Acad. Sci. U.S.A.,* 79, 6285, 1982.
61. **Semancik, J. S. and Harper, K. L.,** Optimal conditions for cell-free synthesis of citrus exocortis viroid and the question of specificity of RNA polymerase activity, *Proc. Natl. Acad. Sci. U.S.A.,* 81, 4429, 1984.
62. **Kiberstis, P. A., Haseloff, J., and Zimmern, D.,** 2′ Phosphomonoester, 3′-5′ phosphodiester bond at a unique site in a circular viral RNA, *EMBO J.,* 4, 817, 1985.
63. **Linthorst, H. J. M. and Kaper, J. M.,** Replication of peanut stunt virus and its associated RNA 5 in cowpea protoplasts, *Virology,* 139, 317, 1984.
64. **Collmer, C. W., Hadidi, A., and Kaper, J. M.,** Nucleotide sequence of the satellite of peanut stunt virus reveals structural homologies with viroids and certain nuclear and mitochondrial introns, *Proc. Natl. Acad. Sci. U.S.A.,* 82, 3110, 1985.
65. **Richards, K. E., Jonard, G., Jacquemond, M., and Lot, H.,** Nucleotide sequence of cucumber mosaic virus-associated RNA 5, *Virology,* 89, 395, 1978.
66. **Collmer, C. W., Tousignant, M. E., and Kaper, J. M.,** Cucumber mosaic virus-associated RNA 5. X. The complete nucleotide sequence of a CARNA 5 incapable of inducing tomato necrosis, *Virology,* 127, 230, 1983.
67. **Collmer, C. W. and Kaper, J. M.,** Double-stranded RNAs of cucumber mosaic virus and its satellite contain an unpaired terminal guanosine: implications for replication, *Virology,* 145, 249, 1985.
68. **Astier-Manifacier, S. and Cornuet, P.,** Purification et poids moléculaire d'une *RNA* polymerase *RNA*-dépendante de *Brassica oleracea* var. Botrytis, *C. R. Acad. Sci. Paris,* 287, 1043, 1978.
69. **Duda, C. T., Zaitlin, M., and Siegel, A.,** In vitro synthesis of double-stranded RNA by an enzyme system isolated from tobacco leaves, *Biochim. Biophys. Acta,* 319, 62, 1973.
70. **Fraenkel-Conrat, H.,** RNA-dependent RNA polymerases of plants, *Proc. Natl. Acad. Sci. U.S.A.,* 80, 422, 1983.
71. **Jaspars, E. M. J., Gill, D. S., and Symons, R. H.,** Viral RNA synthesis by a particulate fraction from cucumber seedlings infected with cucumber mosaic virus, *Virology,* 144, 410, 1985.
72. **Dorssers, L., van der Meer, J., van Kammen, A., and Zabel, P.,** The cowpea mosaic virus RNA replication complex and the host-encoded RNA-dependent RNA polymerase-template complex are functionally different, *Virology,* 125, 155, 1983.
73. **Boege, F., Rohde, W., and Sänger, H. L.,** In vitro transcription of viroid RNA into full length copies by RNA-dependent RNA polymerase from healthy tomato leaf tissue, *Biosci. Rep.,* 2, 185, 1982.
74. **Mühlbach, H.-P. and Sänger, H. L.,** Viroid replication is inhibited by alpha-amanitin, *Nature (London),* 278, 185, 1979.
75. **Sänger, H. L.,** Minimal infectious agents: the viroids, in *The Microbe 1984:* Part I, *Viruses,* Mahy, B. W. J. and Pattison, J. R., Eds., Cambridge University Press, Cambridge, England, 1984, 281.
76. **Solymosy, F. and Kiss, T.,** Viroids and snRNAs, in *Subviral Pathogens of Plants and Animals: Viroids and Prions,* Maramorosch, K. and McKelvey, J. J., Eds., Academic Press, New York, 1985, 183.

77. **Spiesmacher, E., Mühlbach, H.-P., Tabler, M., and Sänger, H. L.,** Synthesis of (+) and (−) RNA molecules of potato spindle tuber viroid (PSTV) in isolated nuclei and its impairment by transcription inhibitors, *Biosci. Rep.*, 5, 251, 1985.
78. **Riesner, D., Henco, K., Rokohl, U., Klotz, G., Kleinschmidt, A. K., Domdey, H., Jank, P., Gross, H. J., and Sänger, H. L.,** Structure and structure formation of viroids, *J. Mol. Biol.*, 133, 85, 1979.
79. **Branch, A. D., Benenfeld, B. J., and Robertson, H. D.,** Ultraviolet light-induced crosslinking reveals a unique region of local tertiary structure in potato spindle tuber viroid and HeLa 5S RNA, *Proc. Natl. Acad. Sci. U.S.A.*, 82, 6590, 1985.
80. **Anderson, J., Delihas, N., Hanas, J. S., and Wu, C.-W.,** 5S RNA structure and interaction with transcription factor A. II. Ribonuclease probe of the 7S particle from *Xenopus laevis* immature oocytes and RNA exchange properties of the 7S particle, *Biochemistry*, 23, 5759, 1984.
81. **Pieler, T. and Erdmann, V. A.,** Isolation and characterization of a 7S RNA particle from mature *Xenopus laevis* oocytes, *FEBS Lett.*, 157, 283, 1983.
82. **Pelham, H. R. B. and Brown, D. D.,** A specific transcription factor that can bind either the 5S RNA gene or 5S RNA, *Proc. Natl. Acad. Sci. U.S.A.*, 77, 4170, 1980.
83. **Honda, B. M. and Roeder, R. G.,** Association of a 5S gene transcription factor with 5S RNA and altered levels of the factor during cell differentiation, *Cell*, 22, 119, 1980.
84. **Shastry, B. S., Honda, B. M., and Roeder, R. G.,** Altered levels of a 5S gene-specific transcription factor (TFIIIA) during oogenesis and embryonic development of *Xenopus laevis*, *J. Biol. Chem.*, 259, 11373, 1984.
85. **Hanas, J. S., Bogenhagen, D. F., and Wu, C.-W.,** Binding of *Xenopus* transcription factor A to 5S RNA and to single-stranded DNA, *Nucleic Acids Res.*, 12, 2745, 1984.
86. **Keese, P. and Symons, R. H.,** Domains in viroids: evidence of intermolecular RNA rearrangements and their contribution to viroid evolution, *Proc. Natl. Acad. Sci. U.S.A.*, 82, 4582, 1985.
87. **Peebles, C. L., Gegenheimer, P., and Abelson, J.,** Precise excision of intervening sequences from precursor tRNAs by a membrane-associated yeast endonuclease, *Cell*, 32, 525, 1983.
88. **Greer, C. L., Peebles, C. L., Gegenheimer, P., and Abelson, J.,** Mechanism of action of a yeast RNA ligase in tRNA splicing, *Cell*, 32, 537, 1983.
89. **Konarska, M., Filipowicz, W., Domdey, H., and Gross, H. J.,** Formation of a 2′ phosphomonoester, 3′,5′-phosphodiester linkage by a novel RNA ligase in wheat germ, *Nature (London)*, 293,112, 1981.
90. **Konarska, M., Filipowicz, W., and Gross, H. F.,** RNA ligation via 2′ phosphomonester, 3′,5′-phosphodiester linkage: requirement of 2′,3′-cyclic phosphate termini and involvement of a 5′ hydroxyl polynucleotide kinase, *Proc. Natl. Acad. Sci. U.S.A.*, 79, 1474, 1982.
91. **Cech, T. R.,** The generality of self-splicing RNA: relation to nuclear mRNA splicing, *Cell*, 44, 207, 1986.
92. **Ruskin, B., Krainer, A. R., Maniatis, T., and Green, M. R.,** Excision of an intact intron as a novel lariat structure during pre-mRNA splicing in vitro, *Cell*, 38, 317, 1984.
93. **Grabowski, P. J., Padgett, R. A., and Sharp, P. A.,** Messenger RNA splicing in vitro: an excised intervening sequence and a potential intermediate, *Cell*, 37, 415, 1984.
94. **Grabowski, P. J., Seiler, S. R., and Sharp, P. A.,** A multicomponent complex is involved in the splicing of messenger RNA precursors, *Cell*, 42, 345, 1985.
95. **Brody, E. and Abelson, J.,** The "spliceosome": yeast pre-messenger RNA associates with a 50S complex in a splicing-dependent reaction, *Science*, 228, 963, 1985.
96. **Peebles, C. L., Perlman, P. S., Mecklenburg, K. L., Petrillo, M. L., Tabor, J. H., Jarrell, K. A., and Cheng, H.-L.,** A self-splicing RNA excises an intron lariat, *Cell*, 44, 213, 1986.
97. **van der Veen, R., Arnberg, A. C., van der Horst, G., Bonen, L., Tabak, H. F., and Grivell, L. A.,** Excised group II introns in yeast mitochondria are lariats and can be formed by self-splicing in vitro, *Cell*, 44, 225, 1986.
98. **Roberts, R. J.,** Intervening sequences excised in vitro, *Nature (London)*, 274, 530, 1978.
99. **Branch, A. D., Shaw, J. F., and Robertson, H. D.,** Competition studies of mild and severe strains of potato spindle tuber viroid co-inoculated into tomato plants, *Virology*, submitted, 1986.
100. **Dickson, E.,** A model for the involvement of viroids in RNA splicing, *Virology*, 115, 216, 1981.
101. **Diener, T. O.,** Are viroids escaped introns?, *Proc. Natl. Acad. Sci. U.S.A.*, 78, 5014, 1981.
102. **Gross, H. J.,** Viroids: their structure and possible origin, in *Subviral Pathogens of Plants and Animals: Viroids and Prions*, Maramorosch, K. and McKelvey, D. D., Eds., Academic Press, New York, 1985, 165.
103. **Semancik, J. S. and Vanderwoude, W. J.,** Exocortis viroid: cytopathic effects at the plasma membrane in association with pathogenic RNA, *Virology*, 69, 719, 1976.
104. **Hari, V.,** Ultrastructure of potato spindle tuber viroid-infected tomato leaf tissue, *Phytopathology*, 70, 385, 1980.
105. **Sänger, H. L.,** Biology, structure, functions and possible origin of viroids, in *Nucleic Acids and Proteins in Plants II*, Parthier, B. and Boulter, D., Eds., Springer-Verlag, Berlin, 1982, 368.

Chapter 3

VIROID PATHOGENESIS AND EXPRESSION OF BIOLOGICAL ACTIVITY

J. S. Semancik and V. Conejero-Tomas

TABLE OF CONTENTS

I. INTRODUCTION

The identification and recognition of small pathogenic RNAs, which have become known as viroids, introduced an unprecedented conceptual extension of our appreciation of pathogenic molecules. So, too, it should not be surprising that the biological expression of these unique molecules would issue from a most intimate relationship with the host. In this encounter between the viroid, which presently defines the lower limit of a minimal infectious entity, and a complex coordinated biological system, we should remain receptive to the revelation of not necessarily an unusual mechanism of expression of genetic potential, but rather a clarification of reactions that are amplified in pathological conditions, but also constitute components in normal growth and differentiation.

The unusual set of properties governing viroid pathogenesis may be clearly reflective of the unique properties of the viroid itself and the relationship with its host. Much of what we shall present is of a highly speculative and correlative nature and is drawn from the limited but diverse observations available, as well as from living with and from viroids for many years. We present these views neither in a presumptive manner nor in the hope of establishing a convincing position, but simply for your consideration as a thoughtful challenge. It remains our perspective that the original contest presented in defining the physical nature of the viroid has been successfully met and the more pressing questions, as well as opportunities, in viroid research reside in the description of the biological activity of the molecule and the mechanism of pathogenesis. Just as the original acceptance of the viroids as a novel grouping of small RNAs demanded a more expansive view of pathogens, the understanding or revelation of the biological interactions between host and viroid requires an intuition coupled with what might even seem outrageous imagination to stimulate as well as coalesce our composition. What we hope becomes clear from this introduction is our purposeful attempt to present not simply a cataloguing of data relevant to the subject of viroid pathogenesis, but to range beyond these constraints. For as much as we must be guided by the disciplined forces of science, biological research can yet be well served by an artistic freedom of thought.

The unique physiocochemical properties of the purified viroid-RNA, including its small size, circularity, and secondary structure resulting in a DNA-like conformation, taken with a nontranslatable "genome" and nuclear locus of concentration, contribute significantly to strict host dependency for replication as well as to the resultant commonality of symptom expression among the different viroids. A major question which still confronts viroid researchers is the rationalization of these intriguing in vitro properties displayed by the isolated viroid-RNA with the in vivo nature of the viroid molecule. Although the viroid has been repeatedly referred to as a "free-" or "naked-" RNA, there exists no data as yet that conclusively establishes such a structure as the biologically active form predominant within plant cells. The special characteristics of viroids and the implicit affinity of host and pathogen in viroid diseases encourage the search for a more direct cause-effect relationship than possible in most biological interactions. Furthermore, these directed studies on viroid-host interaction promise to contribute to our understanding of genome regulation by small RNAs, if the viroid-RNA fulfills its proposed role as a pathogenic form of small nuclear RNA.[1]

The many reviews commenting on viroids have rightfully emphasized the physical properties of the purified RNA, reflecting the bulk of the data available in this area. In the more recent presentations,[1-5] some discussion is usually then given to possible models for replicating viroid-RNA. Pathogenesis is invariably relegated to some brief concluding comments directed to correlations with various small RNAs such as the U-RNA series of mammalian origin, introns, and transposable genetic elements, to provide a capsule speculation of function and origin principally founded in sequence homologies. Although still inadequate to cope with the vagaries of this complex biological system, data in the form of observations, analyses, and responses of viroid-containing plants, tissues, and cell cultures have been accumulated and permit conjecture on the viroid-host interaction as well as on host metabolic responses to viroid infection. Since a virtual flood of secondary process may be activated by a foreign nucleic acid agent at loose in the nucleus, the correlation of these secondary cascades to the primary events of the viroid-host interaction will require time and the exploitation of appropriate experimental systems.

Our approach to deciphering the cryptic message reflected in the varied forms of information related to viroid pathogenesis will be to follow the viroid's path into the plant as the primary infection is established. The dramatic consequences elaborated in the cells and tissues resulting in symptom expression will then be described and, finally, an attempt to correlate suggestions of metabolic aberration as intermediary events in this process will be made. To set the stage, in Figure 1 is presented a composite overview of some of the key factors and structures to which our attention has been focused in viroid-infected cells.

II. VIROID DISEASES AND PRINCIPAL SYMPTOMS

A cataloguing of the known viroid diseases (Table 1) by symptom development reinforces the occurrence of stunting as the primary consequence of the biological expression of viroid-RNA in either the host of original description or the principal assay host. The ten diseases classified within the "PSTV group" display 55 to 80% sequence homology., The term "stunt" even appears prominently in the original description of four of these diseases. Members of the group, such as PSTV, CEV, and TPMV produce a virtually indistinguishable stunting reaction in *Gynura aurantiaca*. The recently described citron variable viroid (CVaV), which induces a reaction in citron identical to CEV when expression occurs, was recovered from a CEV collection and may yet be determined to represent a variant of CEV with a restricted host range resulting from a deletion of 10 to 20% of the nucleotide sequence.[6]

The subgroup reflecting a chlorotic reaction is represented by ASV and CCCV, two viroids with a more distant (10 to 20%) sequence homology to the PSTV group, with the relationship of ChCMV still unknown. Insufficient information is yet available on the *Columnea* viroid or from two brief reports of grapevine viroids to allow positive categorization. However, it is already evident from these reports as well as the reports of the absence of any reaction to a number of viroids of the PSTV group in specific hosts and ASV in avocado that a symptomless condition does occur. The suggestion by Roistacher et al.[7] that the cachexia (xyloporosis) disease, which produces a gumming and stunting in susceptible citrus species, is caused by a viroid, as well as the preliminary report for a viroid-like RNA associated with the apple scar skin disease,[8] must await positive identification of an infectious RNA.

It is important to note that even though a variety of symptoms can be produced, including localized necrosis of vascular tissues, epinasty, rugosity, internode shortening, and fruit and tuber distortion, the predominant occurrence of stunting can be viewed as the most typical expression of viroid infection. The hypersensitive reaction characteristic of the local lesion resulting from many plant virus-host interactions is an extremely rare occurrence in viroid infection, with the combination of PSTV-*Scopolia sinensis* being the best documented case.[9]

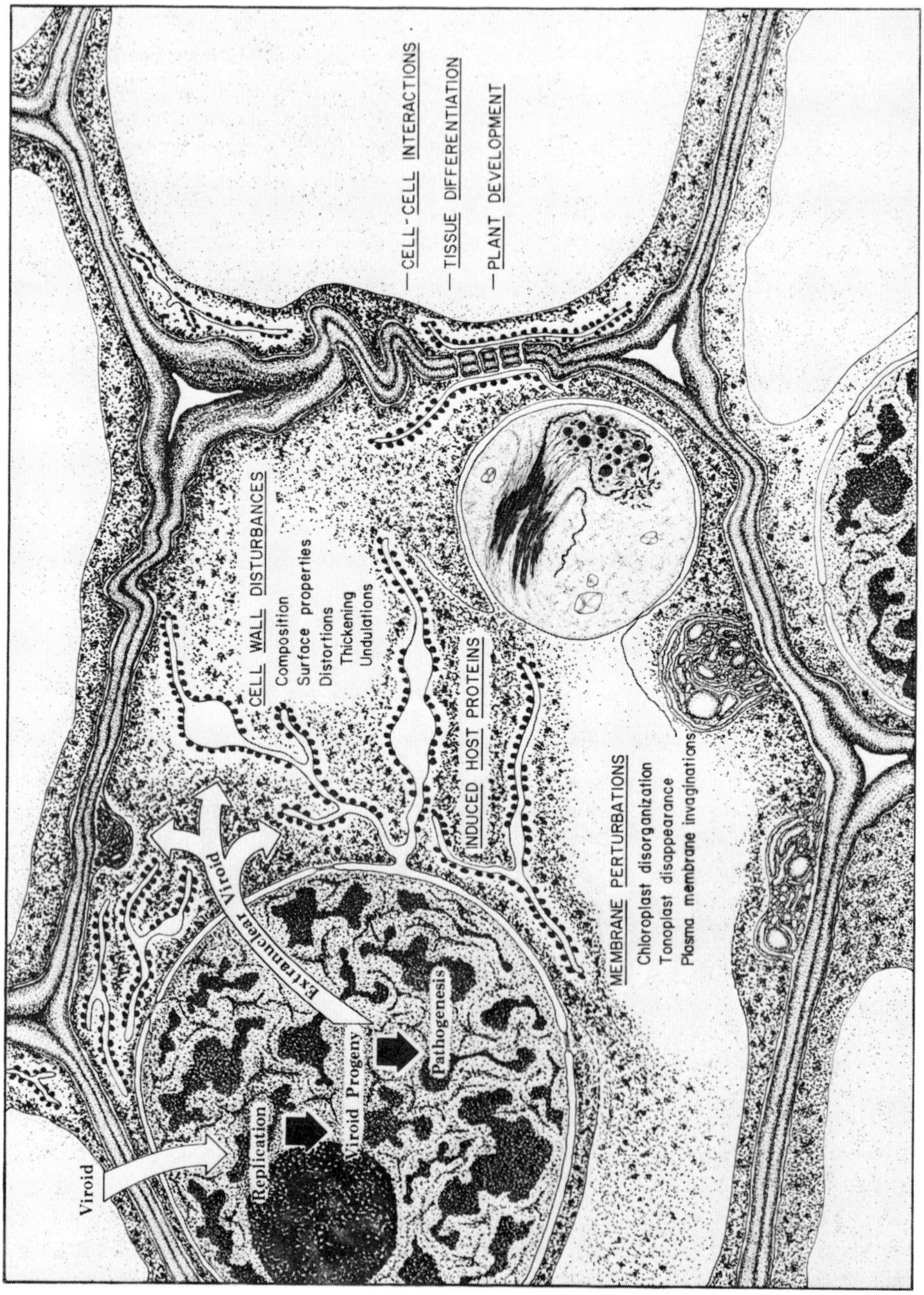

FIGURE 1. Representation of interaction and responses possible in viroid-infected cells.

Table 1
A LISTING OF KNOWN VIROID DISEASES AS CLASSIFIED BY THE PRIMARY EFFECT ON HOST PLANTS

	Ref.
Stunting	
BSV burdock stunt	149
CSV chrysanthemum stunt	150,151
CVaV citron variable	6
CEV citrus exocortis	152,153
CPFV cucumber pale fruit	154
HSV hop stunt	155
PSTV potato spindle tuber	156,157
TASV tomato apical stunt	158
TBTV tomato bunchy top	159
TPMV tomato "planta macho"	50
Chlorosis	
ASV avocado sunblotch	160,161
ChCMV chrysanthemum chlorotic mottle	162
CCCV coconut "cadang-cadang"	163
Uncertain biological expression	
Columnea viroid	164
Grapevine viroids (HSV homology)	165
(GV_f, GV_s)	166
Putative viroid diseases	
Apple scar skin	8
Cachexia (xyloporosis)	7

III. INITIATION OF INFECTION

A. Primary Infection

Critical to our understanding of the early events involved in the establishment of viroid infection is an awareness of (1) the physical nature of viroid-RNA as it influences (2) the rate of movement of viroid in infected plant tissues. Implicit in this consideration is the realization that the form of the viroid-RNA in the plant may change as reflective of discrete levels of association with host components and establishment of centers of viroid-host complexes involved in replication and biological activity.

Since primary infection in viroid diseases usually involves a systemic response in newly developing leaves initiated at some distance apical to the site of administration of the inoculum, the importance of vascular tissues terminating at some susceptible or "receptive" cell types has been proposed.[10,11] Even though leaf rubbing is effective in inoculation of herbaceous hosts, in practical application some form of stem slashing with razor blades or pricking with needles has proven most effective as a general inoculation procedure. In this procedure, the viroid inoculum is deposited in direct contact with vascular tissues. The initial plant response of aberrations in the newly developing leaves in the form of chlorosis, vein-banding, and epinasty does not occur for 10 to 14 days at the earliest. In this process, fully mature leaves as well as those in the final stages of leaf expansion are apparently by-passed, either because of (1) the outflow of translocation from these leaves, (2) the absence of "receptive" cells, or (3) the inability of the viroid-RNA to reach these cells. It is uncertain whether the viroid inoculum is translocated from the initial point of slashing by either the xylem or phloem components of the vascular tissues or both. Since the site of primary infection occurs in immature leaves, we may postulate that the relatively stable viroid inoculum can be transported by xylem to the growing apex where it is brought in contact with viable "target" cells in the region of differentiation of the procambial initials. The more rapid rate of cell division at the site of initiation of procambial strands[12] may be significant to the successful survival of the viroid-RNA.

The directional specificity of this inoculum flow is demonstrated by infection of CEV in *Gynura aurantiaca* (Figure 2E) on occasions when only one half of the first leaf that displays symptoms is affected.[11] It would seem that the viroid-RNA is introduced into only a portion of the vascular ring of the stem, and lateral spread is minimal when compared to the mass vertical flow. The half-leaf infection model suggests that the inoculum can be carried in the vascular tissues which connect from the main vascular elements of the stem to only one side of the petiole and midrib and, subsequently, the secondary veins of the half-leaf so served. The inoculum level which either reaches or survives in mature asymptomatic leaves is insufficient to be detected by polyacrylamide gel electrophoresis (PAGE) analysis or bioassay. Furthermore, viroid replication is not detected in callus cultures generated from these symptomless mature leaves.

B. Association with Centers of Mitotic Activity

Although the suggestion has previously been developed for a direct relationship between viroid synthesis and mitotically active tissues,[1,13] it has not been possible to demonstrate that mitosis is essential for viroid replication. In fact, evidence to the contrary exists. When leaf discs are taken from systemically infected leaves and held in culture on agar for as long as 1 month, viroid replication continues as long as tissue viability is maintained. Also, synthesis of both PSTV and CEV is evident in tomato cell suspension cultures in the later stationary phase of the growth curve in which cell division has diminished. Successful inoculation of isolated protoplasts by CPFV and PSTV[14,15] can be accepted as additional evidence that conditions can be effected to accomplish infection of mature cells.

Nevertheless, the importance of "developmentally juvenile" target cells in the establishment of the primary infection should not be dismissed. Although an exact cell stage or active biochemical process of the cell governing successful viroid infection cannot be indicated at present, the simple availability in the region of dividing cells of diverse cell types reflecting unique physiological conditions may present a more favorable locus for the survival and colonization of viroid-RNA. It may follow that as the bulk of the cell mass undergoes vacuolation and assumes a nonmitotic G_0 state of the cell cycle, the cellular conditions no longer favor acceptance of a newly introduced foreign RNA, even though the capacity to replicate viroid molecules previously established is maintained.

This association between actively dividing cells and the *initiation* of viroid infection is presented simply for consideration and to raise the questions if and why this should be so. The observation has often been presented that the gross response of a plant to viroid infection resembles a hormonal imbalance. This very dramatic clue can be coupled with the understanding that the cell cycle and cell differentiation are directed by hormonal controls.[16] Therefore, the viroid affected cells may be acted upon very early in their developmental progression, long before the more massive systemic responses are displayed at the whole plant level. Specific interactions with "immature" cells may not only assure replication of the viroid-RNA but also direct what we view as the pathogenic reaction.

Since the G_0 state has been defined as a phase in which specific biochemical functions can be expressed quite apart from the proliferation cycle,[17] it may be implicated in a metabolic segregation between receptive and nonreceptive cells. In the absence of critical experimental approaches, this suggestion must, of course, be viewed as simply speculation. Nevertheless, it has been generally acknowledged that the regulation of cell division and development and the accompanying specific biochemical condition is governed by a cytokinin gradient in balance with an auxin gradient. If the viroid-RNA can be considered as a specialized class of regulatory snRNA, its persistence as well as functional expression may depend upon a critical interfacing with a specific cell developmental stage in which similar host snRNAs are active. The conclusion to this scenario envisions the affected host's subsequent inability to adequately control either the synthesis or the activity of the novel viroid-snRNA, which results in cells set inexorably on a path of abnormal development. This entire process, of course, implies a close degree of coordination between nuclear activities and cell growth. More specific suggestions on the mechanism of phytohormone action will be presented when possible molecular events of viroid infection are discussed. Finally, since vascular differentiation is initiated with the formation of the procambium, again under the control of phytohormones, we can propose the mixing of the critical ingredients that results in ultimate disease expression: the infectious RNA, the target cell, and the active intermediary metabolic processes of the host.

Further evidence for an association or affinity of viroid with centers of active cell division can be drawn from diverse experimental observations. As previously mentioned, the majority of viroid diseases are expressions of the direct affect of the pathogen on apical growth and differentiation. Not only does the apex become the principal initial site of viroid replication, but secondary spread is manifest in all subsequently developed lateral buds, as well as in root tips. Even in those specialized cases in which stunting is not the predominant expression of viroid disease, such as the severe chlorosis of avocado leaves by ASV, the pattern of the severely affected tissues resembles the boundaries of a marginal leaf meristem. Furthermore, the accumulation and replication of ASV appears to be restricted to symptomatic tissues and delimited by the non-necrotic boundary between the normal-appearing green tissue and the severely chlorotic region. The condition enabling symptom expression in the avocado leaf might be analogous to the relationship that exists in the apices of hosts of other viroids between the viroid and "target" cells, with the target cells laid down in the ASV case principally by the marginal meristem. Viroid-RNA produced in the chlorotic zone might not become established in the green areas if these cells have become "developmentally resistant" to viroid infection. The high concentration of ASV that accumulates in the avocado fruit might be related to the unusual frequency of cell division in avocado as a major factor in fruit growth throughout its development on the tree.[18]

The rather unusual expression of CEV infection of the *Poncirus trifoliata* rootstock results in bark shelling or scaling immediately below the scion-rootstock union. The region directly above the union has been described as displaying the properties of juvenile tissue and as a center of active cell division. Also, increased meristematic activity has been observed[19] in midvein tissues of citron *(Citrus medica),* an indicator host for CEV. When tumors were induced by *Agrobacterium tumefaciens* in stems of tomatoes infected with CEV, the concentration of viroid in the cells proliferating from the tumors was significantly greater than in the nontumorous stems,[13] indicating that the neoplastic cells provided an excellent tissue for viroid synthesis. In fact, suspension cultures of tomato cells containing viroid-RNA have maintained a stable titer of viroid for several years.[20]

The shoot tip culture and shoot tip grafting procedures[21,22] which have been widely utilized to free plant tissues from plant viruses and viroids may be cited as evidence for the absence of a functional relationship between viroids and meristematic centers. Although it was reported by Stace-Smith and Mellor[23] that it was easier to free plant tissues of viruses than viroids, which might indicate the latter to be a more invasive agent, the use of themotherapy in conjunction with shoot tip culturing procedures probably enhanced the replication and spread of the viroid. More recent studies[24] which employed cold treatment in the successful application of these procedures to free plants of PSTV support this contention. Furthermore, in practical applications, CEV and the agent of the xyloporosis disease of citrus are the easiest to eliminate from plants harboring mixtures of viruses and viroids.

The successful recovery of healthy tissue from an infected shoot tip is usually accomplished when the excised apex is less than 0.2 mm in length. The fact that such tissue is apical to the development of any vascular tissue dictates that the presence of any viroid-RNA must be a result of cell-to-cell movement. Such restricted movement would tend to mitigate against the physical contact of the most apical cells with the viroid. Furthermore, if during cell-to-cell spread, the movement of the viroid[25] is not as a free RNA molecule but rather complexed with host protein in a viroid ribonucleoprotein particle (RNP), cell-to-cell movement may be quite slow and may not keep pace with cell division. Alternatively, cells in the apex containing a viable replicating center may not produce an "export" quality or quantity of viroid-RNA until later in differentiation.

C. Secondary Infection

Spread of progeny viroid from the locus of primary infection can be affected by the in vivo nature of the infectious agent. What was introduced to the plant as a highly purified RNA, although replicated with due fidelity, probably becomes associated with host components (enzymes, membranes, etc.) during the process of replication and may perpetuate these associations as a means of survival, spread, and expression of biological activity. It is worth reiterating here that little data are available describing the in vivo properties of the viroid and supporting the free- or naked-RNA structures much lauded in discussions about viroids. Configurations potentially involved in viroid transport include dsRNA, viroid-complementary structures, and viroid-host complexes. Such complexes may reflect association with proteins, as in a viroid-RNP, as well as membrane associations as might occur in a replicating system.

An interesting study by Kryczynski[25] can be used to compare transport of viral nucleoprotein and free- and viroid-RNA inocula. By employing a tissue implant inoculation technique it was possible to monitor the efficiency of transmission and rate of translocation from cell-to-cell in the absence of vascular connections. The percent transmission from tissue infected with a viroid (PSTV) could be equaled by a nucleoprotein virus (tobacco rattle virus-stable form) and was only slightly better than a free-RNA viral infection (tobacco rattle virus-unstable form). This suggests that the predominance of the primary infection molecule as a 48×10^6 dalton nucleoprotein, a 2×10^6 dalton RNA, or a 1.2×10^5 dalton RNA had little effect on the efficiency of transmission. The transmission of various potato viruses was, however, considerably less. For instance, potato virus M demonstrated only a 56% efficiency of transmission compared to 95% by PSTV. Nevertheless, when the rate of translocation from the infected tissue implant to the uninoculated host for potato virus M was compared to that for PSTV, no differences between the rate of movement of the viroid and the virus were observed. This data might be interpreted as indicating that the cell-to-cell spread was not dependent upon the morphology of the parental infection unit, but that these diversely sized infectious RNAs may have been transported as similar RNP-like complexes.

Subnuclear fractionation of PSTV-infected tissues has demonstrated the association of viroid-RNA with pelleted nucleoli.[26] Since the viroid could be released from the nucleolar pellets by high salt or Mg^{2+} treatment, it was suggested that the viroid exists as an RNP in association with the nucleolus. Since we do not yet know if exogenously added purified viroid-RNA might form nonspecific ionic interactions with nucleoli, these data remain difficult to interpret. Nevertheless, since most snRNAs exist as snRNPs, the putative viroid-RNP remains an attractive structure for an intracellular viroid form and it would seem that this structure would be a prime candidate for cell-to-cell spread from the primary infection site.

Seed transmission, presumably by viroid-containing pollen, can be viewed as a specialized type of secondary infection since the contaminated seed are derived from inoculated plants. For the most part, seed transmission of viroids is not common, being principally reported for PSTV in potatoes[27] and *Scopolia*[28] and ASV in avocado.[29] In citrus, viroid-infected seedlings have never been reported, even though extensive surveys have been made in conjunction with sanitation programs. In the particular case of CEV on severely affected citron, the few seeds produced are rendered sterile or aborted.

The symptomless carrier condition whereby ASV is carried in avocado seed from infected plants and is totally nonexpressing in progeny trees introduces a most intriguing host-viroid interaction. It would seem that the otherwise susceptible host has acquired the ability to accept the viroid-RNA as a nonthreatening agent. The embryonic shoot meristem carries the viroid. Could it be that the viroid-RNA is, in effect, ''imprinted'' onto the host RNA metabolism when cell division is resumed upon germination and is then synthesized as a ''normal'' component of the nucleic acid profile? It would seem that ''resistance'' to the clinical expression of infection can be acquired by the exposure to viroid at a sufficiently early stage of development of the host to result in the suppression of pathogenesis.

The initiation of viroid infection, then, establishes some subtle but characteristic properties of this group of pathogens. The polarity demonstrated by the viroid-RNA inoculum for some still ill-defined ''target'' cells suggests the necessity for developmentally receptive cells. The subsequent systemic spread and association of viroid with meristematic centers suggests an almost cooperative balance which can result in the total acceptance of the foreign infectious RNA. In most cases, however, deleterious effects result which may reflect an imbalance in normal growth processes or, in effect, a ''developmental disease.''

IV. SYMPTOM DEVELOPMENT

A. Apparent Manifestation of Viroid Infection

What has been presented in the foregoing section is an attempt to describe a commonality of symptom expression among the various viroid diseases. We would now like to reconcile some of the distinctly different variations in symptom expression to a commonality of origin of host-viroid interactions. In addition to the principal symptom of stunting, which is accompanied by abnormal leaf development resulting in epinasty and rugosity, localized regions of necrosis, especially in the midvein, are often observed in association with the leaf curling symptom. Fruit or tuber distortions and reduced size are also evident in CPFV in cucumber, ASV in avocado, CEV in citron, PSTV in potato, HSV in hops, CCCV in coconut, and a wide range of viroids in tomato.

These fruit and leaf symptoms can be accompanied by chlorotic zones of irregular dimensions. Some such zones are hardly visible while others may be highly evident as on chrysanthemum with ChCMV. Most striking displays of chlorotic symptoms occur on coconut with CCCV and avocado with ASV. No stunting is observed in either of these host-viroid combinations. Furthermore, both of these viroids appear to be highly restricted in host range. It is reasonable to speculate that the limited genome of CCCV and ASV, only 70% of the PSTV group, may introduce a factor in the distinct symptom expression and almost single host exclusivity.

Perhaps the most variable expressions of a single viroid in concert with different hosts are seen with the citrus exocortis viroid. The original description of the disease in citrus[30] describes the bark shelling or scaling effect (Figure 2A) on a sensitive rootstock such as *Poncirus trifoliata* when budded with an infected scion. When transmitted to citron *(Citrus medica),* the principal indexing host for citrus, epinasty, midvein callus deposits, and dwarfing are evident (Figure 2D,E).

These responses in different hosts can be viewed as a continuum of severity of symptoms involving similar tissues and/or the ability of the viroid to more completely colonize host tissues. Foliar symptom expressions may be due to effects on the differentiation process of cells derived from the apical meristem, while bark cracking probably results from effects on cells derived from the lateral meristems. Both reactions, nevertheless, are closely associated with cells in proximity to vascular tissues. The bark scaling reaction on susceptible

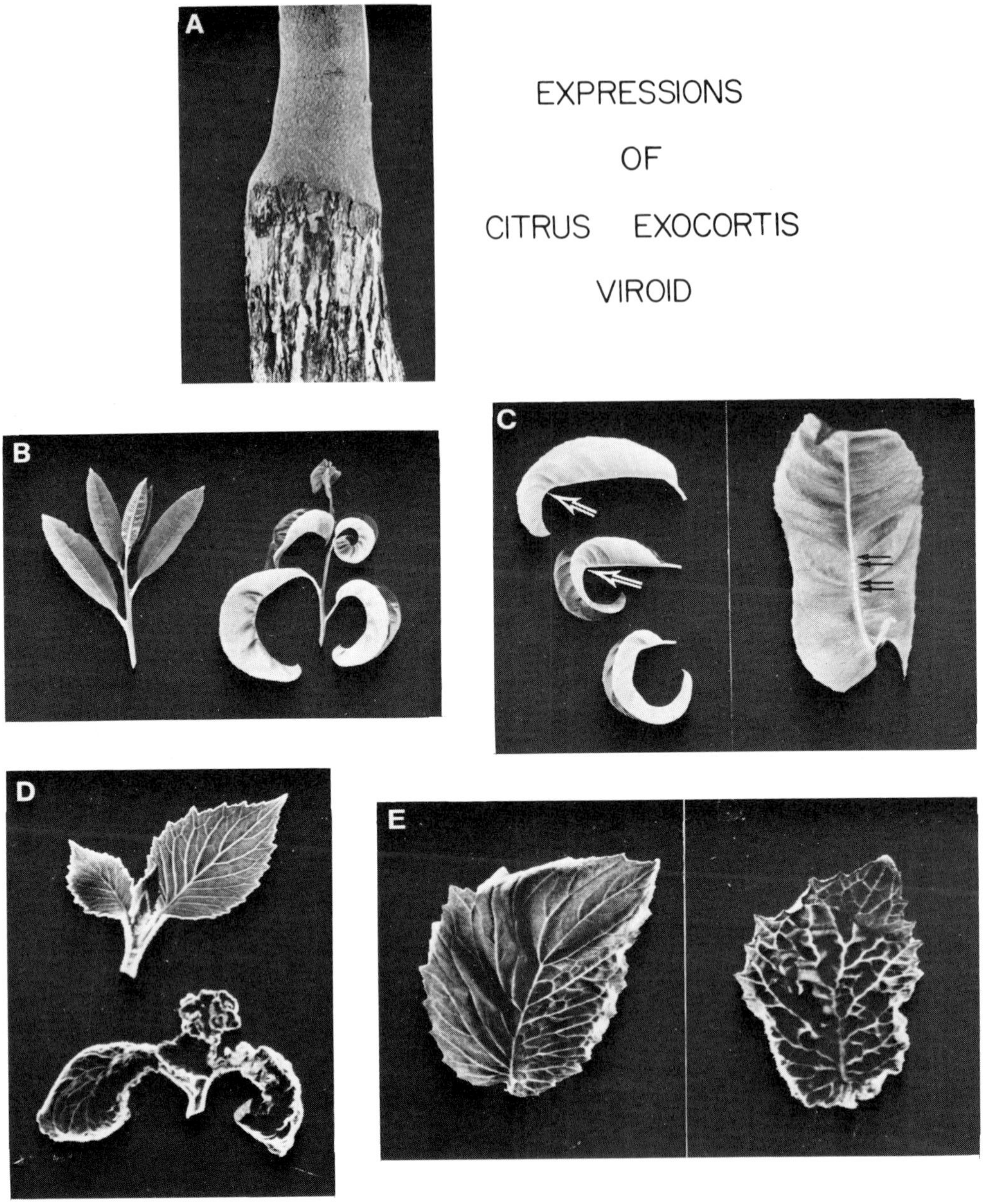

FIGURE 2. Symptom expression of citrus exocortis viroid (CEV) on (A) *Citrus limon* var. 'Eureka' scion on 'Morton' citrange rootstock displaying severe bark scaling, (B) healthy (left), and CEV-infected (right) *Citron medica* with leaf epinasty caused by localized areas (arrows) of necrosis on the lower surface of the midrib as in (C), and (D) healthy (upper) and CEV-infected (lower) *Gynura aurantiaca* displaying severe epinasty and rugosity. Half-leaf partitioning of asymptomatic (left half) and symptomatic (right half) of first affected leaf (left) and a systemically infected leaf (right) with severe vein clearing (E).

rootstock begins immediately below the original budunion with the CEV-containing scion from which the viroid-RNA is translocated. It can be conjectured that the juvenile tissues and the center of meristematic activity of the scion located immediately above the budunion contribute to this apparent viroid flow. It is possible to detect infectious viroid in both scion and rootstock in the region of the bud-union by transmission from excised tissue to citron. Lesions in the inner bark of CEV-infected *P. trifoliate* display hypertrophy and hyperplasia in the ray cells, sieve tubes, and phloem parenchyma, in addition to necrotic areas which presumably result in the sloughing of bark fragments.[31] Histochemical tests also indicate the presence of wound gum in the affected cells.[32]

Epinasty of the tip leaves of CEV-infected citron occurs as a result of localized necrosis in cortical tissue on the underside of the midrib (Figure 2C).[33] Corky lesions or vertical cracking of the stem are also evident.[34] The diseased cortex, especially near the phloem, contains a greater number of small parenchyma cells than the healthy cortex. As the lesion becomes more evident, cell division occurs around the affected parts, with cells in the phloem and the cortex becoming necrotic. Leaf epinasty resulting from normal growth of the upper surface precedes this reaction. Although the viroid was translocated to leaf tissue to effect the diagnostic symptom, as in the case of the scaling reaction observed in *P. trifoliata*, vascular tissues and cells in close proximity appear most severely affected. Perhaps a hypersensitive response in these regions actually limits the spread of the viroid and, at the same time, restricts internode elongation.

The reaction to CEV in *Gynura aurantiaca* does not appear to be as severely delimited as in *Poncirus* or *Citrus* nor are centers of localized necrosis on the leaf midveins as evident. As a result, the secondary veins also display an epinasty reaction, giving the entire leaf surface a rugose texture. The more complete distribution of the viroid throughout the leaf lamina is also reflected in small irregular chlorotic areas of new leaf growth.

What can, therefore, be proposed in this series of host reactions is a progression in the invasiveness of the viroid into vascular tissues, which is seemingly limited by the host response. This progression ranges from the localized hypersensitive syndrome which occurs in *P. trifoliate* to a systemic reaction limited to only the major veins of citron, but which affects the secondary vascular tissue and the leaf lamina of *G. aurantiaca.* Although at first appearance these reactions may appear quite distinct, the effect of viroid infection in these hosts from different gene pools representing the familiar *Rutaceae* and *Asteraceae* is, hence, quite similar. Since the reaction on tomato, potato, and petunia[35] is identical to that on *Gynura,* this range can also be broadened to include the *Solanaceae.* This observation encourages one to consider a specific response of the host to viroid infection which may even be a most direct cause-effect relationship unparalleled in interactions between more complex organisms. The consistent occurrence of viroid effects on vascular tissues in association with meristematic centers directs our attention to early developmental events of cell differentiation. The obvious similarity of host response to the presence of viroid and to exposure to excess ethylene suggests involvement of hormonal elements. This recurrent theme will reemerge in this chapter.

Relative to disease development, an interesting relationship has been proposed between CEV and a smaller (105,000 dalton) viroid-RNA recently described by Schlemmer et al.,[6] citron variable viroid (CVaV), which had been included in an exocortis disease collection for many years. CVaV was originally derived from 'Eureka' lemon (*Citrus limon* (L.) Burm.) with severe bark scaling. However, when transmitted to 'Calamondin' *(Citrus madurensis Lour.)* on 'Morton' citrange (*Poncirus trifoliata* (L.) Raf. x *Citrus sinensis* (L.) Osb.) rootstock, only mild bark scaling was evident. The symptoms in citron were identical to those caused by CEV, but were, however, irregular in expression, with symptomatic shoots interrupted by flushes of normal-appearing growth. Symptom expression was apparently influenced by light quality or duration. No interference or protection occurred between CEV and CVaV when both were maintained in citron. These data, taken with the 10 to 15% variation in nucleotide number, suggest that the two viroids are distinct. Furthermore, the CVaV host range is presently restricted to *Citrus* species. Nevertheless, because of the uncertainty of cross-protection responses in viroid-infection of citrus, it is possible that CVaV represents a deletion variant of CEV. If sequence homology determines this to be true, the relationship between the viroid genome and host response in citron can be more specifically correlated. Furthermore, the temporal relationship between the synthesis of viroid, the initiation of new growth, and the induction of pathogenesis are dramatically represented in this system.

Extending this relationship between a limited viroid genome and symptom expression, both CCCV and ASV are composed of only about 247 nucleotides or are about 70% of the size of viroids of the PSTV group. Both are characterized by a chlorotic reaction on the stems, fruit, and leaves and are extremely restricted in host range, principally to coconut palm or avocado, respectively. In these diseases, the most apparent manifestations of viroid infection as well as viroid accumulation[36] are associated with chloroplasts and plastid development. Perhaps this is an indication of a preference of the viroid for the chloroplast over the nucleus as a receptive autonomous genetic system. On the basis of this affinity for plastids as well as the limited sequence homology to the other viroids, the interaction of CCCV and ASV with their respective hosts must be viewed as a subset within the general class of host-viroid interactions.

In summary, it becomes apparent in considering even these gross expressions of viroid-host interactions that the pathological response of the host is centered in a vascular locus. Environmental constraints and host cell development, including a temporal aspect, in concert with viroid presence as governed either by its synthesis or translocation combine to contribute to the nature and extent of the host response.

B. Tissue Morphology

Limited studies have been made on the tissue morphology of viroid-infected plants. Most cytological observations have focused at the ultrastructural level on the condition of organelles in symptom-expressing foliar tissues. In the discussion of the effects of CEV on various hosts, we have already presented an anatomical interpretation of the epinasty response which results from viroid affects on cortical tissues of the leaf midrib in citron. Implicit in the process of the generalized stunting seen in many viroid-host interactions is the retardation of internode elongation, perhaps resulting from interference with vascular development or tissues directly associated with conductive elements.

Light microscopic examinations of transverse sections of leaf tissues have been made on HSV-infected hops and cucumber as well as CEV-infected tomato.[37] In these three host-viroid combinations a similar lack of regular cell arrangement in the leaf laminae was observed, with the most pronounced distortion occurring in mesophyll and epidermal cells. When healthy and CEV-infected tomato leaf discs were placed on nutrient agar and held in culture for 1 week to 1 month, the distinction between the tissues became even more pronounced.[38] Healthy discs expanded to more than twice their initial dimensions while the infected discs enlarged only slightly. Most striking in the latter tissue was the almost total absence of intercellular spaces in the spongy mesophyll regions and the severe disorganization of vascular bundles. A similar reduction in intercellular spaces and cell size was noted in chlorotic tissue of CSV-infected chrysanthemum.[39] This would point to some impairment in the loosening of contact between cells required for the enlargement of cells and for adequate separation to permit aeration of the tissues. An increase in cell-cell adhesion might, therefore, occur as a result of viroid infection.

This suggested relationship of alterations in cell wall properties to viroid infection was supported in a study by Momma and Takahashi.[40] By examining cytopathic changes in the shoot apex of hop plants infected with HSV, a direct correlation between the detection of viroid and primary cell wall lesions in the form of thickened or undulated cell walls was possible. The apical 0 to 0.2 mm portion of the shoot was found to be HSV-free as determined by bioassay of excised fragments. No cytopathic alterations were detected in the organelles of cells of the 0.2-mm long shoot tip which consisted of the tunica, corpus, ribmeristem, first leaf primordium, and first stipule primordium. Successive segments of 0.2 to 0.5 mm and 0.5 to 1.0 mm displayed increasing amounts of HSV. Aberrations in the cell wall structure, which occurred only in infected plants and consisted of undulations and irregular thickening, were first observed in the third leaf primordium which developed from the 0.2 to 0.5 mm region of the apex. Furthermore, viroid-free plantlets could only be recovered when extremely small (less than 0.3 mm) shoot tips were cultured in vitro. It should be recalled that such an excised region would probably exclude the procambial initials from which vascular tissues are derived. The presence of viroid within 0.2 to 0.5 mm of the shoot apex of the host offers an explanation for the difficulty reported in eradication of PSTV from potato by culture of 0.5 to 1.0 mm bud apices.[23] Cell wall structure in the corresponding regions of shoot apex of plants freed of HSV by tip culture proved to be normal in development. No additional morphological effects were noted in these studies; however, development of chloroplasts was not evident in the examined area of the shoot apex.

Leaf and cell surface characteristics have been studied with the HSV-hop and CEV-tomato systems, respectively. Epidermal cells of infected hops did not possess a cuticular surface layer composed of fold-like structures as observed on the healthy leaf surface.[37] Suspension cultures of CEV-containing tomato cells were characterized by protuberances of unknown composition which were not found in scanning electron micrographs of non-CEV-containing cells.[41] These preliminary observations may present evidence for the accumulation or exudation of compounds which may be responsible for altered cell surface properties such as the apparent adhesion of viroid-infected cells which results in reduced aeration of leaf tissues.

V. CELLULAR CYTOPATHOLOGY

Included in the various ultrastructural studies of viroid diseased tissues are six viroids in about 11 different host-viroid combinations. A summary of the cell structures that appear to be affected by the presence of viroid is presented in Table 2. Most tissues inspected were taken from systemically infected plants. Only examination of CSV on chrysanthemum[39] and PSTV on *Scopolia*[9] included tissue from chlortic or necrotic local lesions, respectively. Since the extent of viroid invasiveness into various tissues has not been determined for each viroid-host combination, the possibility of localization of viroid-containing cells in infected plants cannot be excluded. Therefore, the selection of appropriate tissues in infected plants as well as developmentally or chronologically similar tissue from healthy plants as control samples are critical to these analyses. For the most part, control specimen tissues have been taken from healthy plants of comparable age. In some cases where clearly defined affected regions could be discerned, such as in the chlorotic lesions on CSV-infected chrysanthemum,[39] the extensive chlorotic or "bleached" zones on ASV-infected avocado,[42] and the "half-leaf" symptom on CEV-infected *Gynura,*[11] sampling was made of the apparently normal and affected regions. The detection of viroid-RNA could be directly correlated with the presence of symptoms in some of these tissues.[11,42,43]

Further complicating the interpretations from these studies is the absence in infected plants of any affect(s) known to be exclusively viroid induced. Many of the observed host responses to the presence of the viroid resemble responses which result from either virus infection or physiological trauma, such as stress.

Nevertheless, as a result of the survey presented in Table 2, our attention is directed to aberrations in cell wall structure and chloroplast organization as possible primary sites of viroid-induced lesions. In addition, various reports suggest that other membranes such as the plasmalemma and tonoplast may also display effects of viroid infection. It is interesting to note that even though the nucleus has been generally accepted as the site of viroid replication and potentially the primary events involved in pathogenesis, a unanimous consensus emerges which supports the absence of any cytopathic effect on nuclear structure.

Table 2
CELL STRUCTURES AFFECTED BY VIROID INFECTION

Viroid/host	Cell wall	Plasma membrane	Chloroplast	Endoplasmic reticulum	Tonoplast	Mitochondria	Ref.
ASV/avocado	−	+	+	−	−	−	42
CEV/*Gynura*	−	+	−	−	−	−	11
Gynura	+	+	+	−	−	−	44
tomato	+	−	+	−	−	−	37
tomato	+	+	−	−	−	−	20
tomato	+	−	+	−	+	−	19
CPFV/cucumber	+	−	+	−	+	−	19
tomato (symptomless)	−	−	+	−	+	−	19
CSV/chrysanthemum	−	−	+	−	−	+	39
HSV/hops	+	−	+	−	−	−	37
cucumber	+	−	+	−	−	−	37
cucumber	+	−	+	−	+	−	19
tomato (symptomless)	−	−	+	−	+	−	19
PSTV/tomato	−	+	+	−	+	−	46
tomato	+	−	+	−	+	−	19
Scopolia (local lesion systemic)	+	−	+	+	+	+	9

A. Cell Wall

A clear correlation between the detection of infectious HSV and the abnormal development of the primary cell walls in shoot apex tissue has already been introduced (Figure 3).[40] The appearance of undulating cell walls with regions of irregular thickening was not accompanied in this case by aberrations of the plasmalemmae or membranes of other cell organelles. Interestingly, the tissues sampled were not grossly affected by the presence of the viroid. Similar observations have been made in a number of other viroid-host combinations in foliar tissues of severely affected tissues. In two additional cases in which viroid infection did not result in any externally visible symptom expression, CPFV- and HSV-infected tomato, no changes in the cell wall structure were observed.[19] Cell walls of CEV-infected *Gynura aurantiaca* have been described as "corrugated" with an occasional undulating broad band in the area of the middle lamella in cells where the cell wall was not distorted.[44] Whether these observations suggest that abnormal cell wall development in viroid-infected tissues is a response to inhibition of cell-cell loosening caused by increased adhesion and/or fundamental lesion(s) in cell wall chemical composition remains to be determined. Nevertheless, with the pronounced symptoms of stunting resulting from interference in internode elongation and expansive cell growth expressed in viroid diseases, the existence of cell wall lesions would seem a conceptually compatible observation.

Less typical cytopathological alterations in the cell wall have been observed in PSTV-infected *Scopolia sinensis*.[9] Infection in this system is characterized by a local lesion response at about 10 days postinoculation and a systemic chlorosis at about 22 days postinoculation, followed by systemic necrosis. Wall deposits between the plasmalemma and the cell wall were localized in tissues adjoining local lesions as well as in systemically infected tissues. Since these structures resembled the effect noted in necrotic local lesion development in potato virus M-infected bean,[45] it would seem that the response in PSTV-affected *Scopolia* is distinct from those of other viroid-infected tissues and more reflective of the hypersensitive reaction.

B. Plasma Membrane

The difficulty in establishing the existence of a cytopathic effect is clearly exemplified in the question of whether viroid induces alterations of the plasma membrane or of paramural

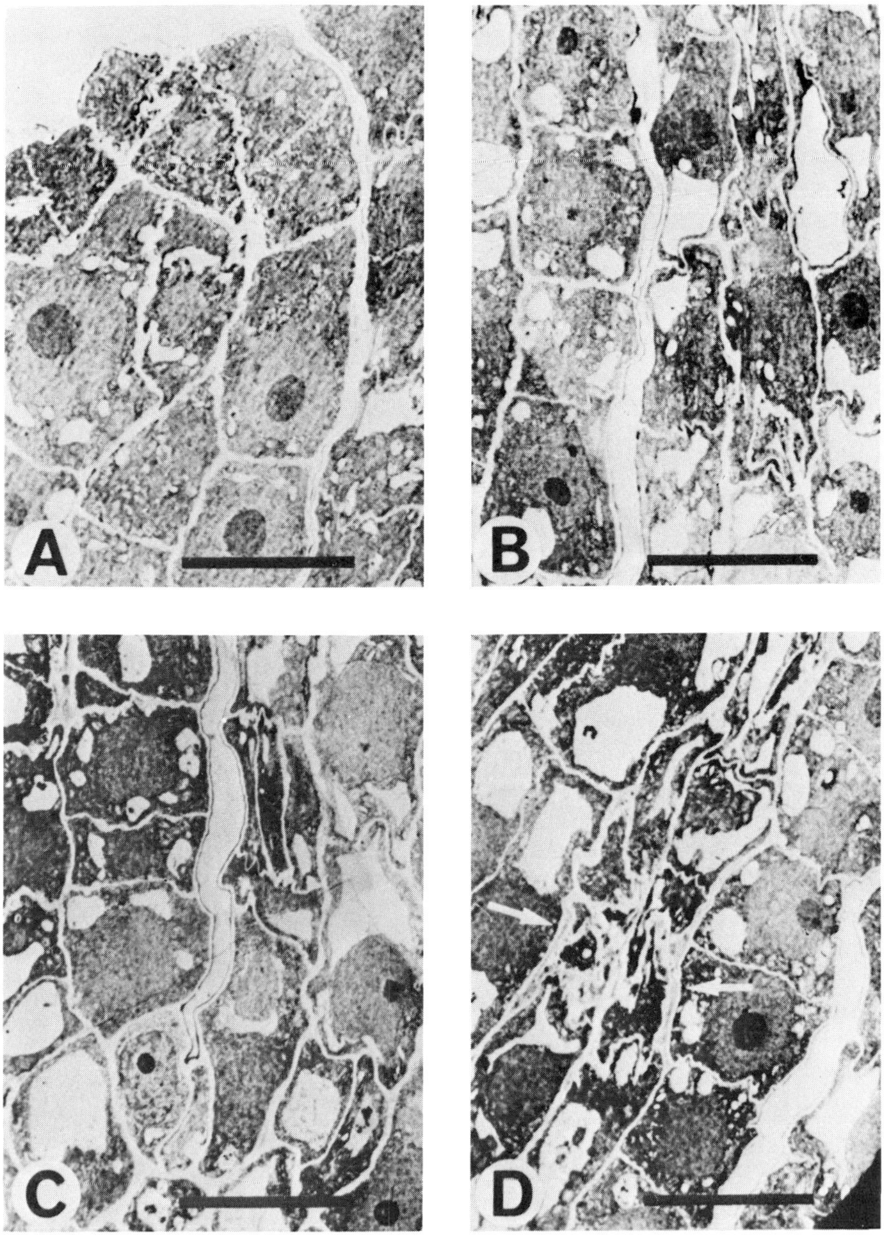

FIGURE 3. Transmission electron micrographs of the shoot tip of a hop stunt viroid-infected hop plant. Note the undulations of cell walls in the ventral (A, B, C) side and central part (D) of third leaf primordium. Arrows in D indicate the irregularity in thickness of cell walls. Bar represents 10 μm. (From Momma, T. and Takahashi, T. *Phytopathol. Z.*, 106, 272, 1983. With permission.)

bodies in the cell wall-plasma membrane zone (Figure 4). In tissues affected by CEV and PSTV, reports suggesting both the presence[11,44,46] and absence[19] of membrane proliferations have been made. These reports involve both increased frequency of plasma membrane proliferations termed ''plasmalemmasomes'' by Semancik and Vanderwoude,[11] and altered internal structure reported by Wahn et al.[44] The perception of a subcellular alteration as potentially subtle as plasma membrane distortions is particularly dependent upon the judicious

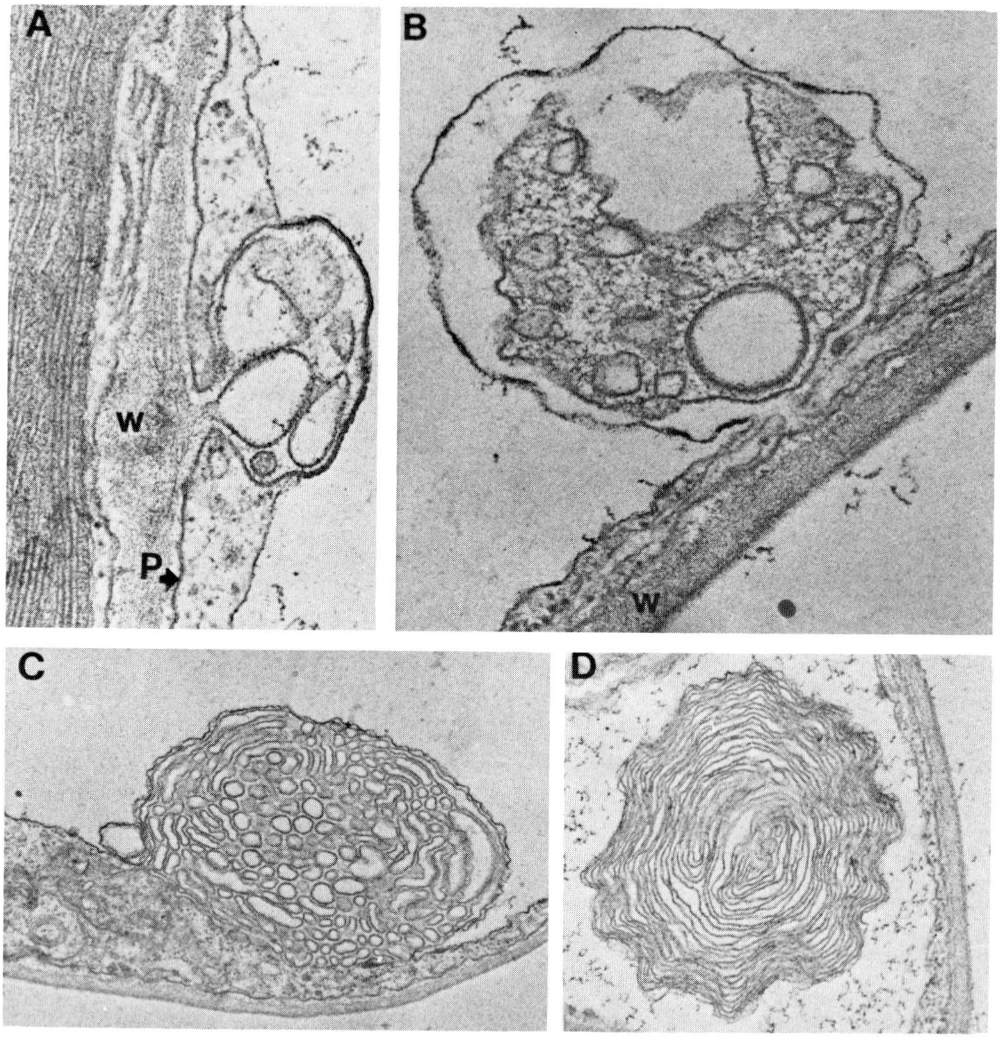

FIGURE 4. Paramural bodies and membrane abnormalities in potato spindle tuber viroid-infected tomato leaf cells. (A) Plasmalemmasome formed by invagination of plasma membrane into vacuole, (B) similar body formed by tonoplast, (C) multiple vesiculated paramural body, (D) watchspring-like multilayered membrane body found in cytoplasm. W = cell wall; P = plasma membrane. A to C, magnification × 50,500; D, magnification × 27,500. (From Hari, V., *Phytopathology*, 70, 385, 1980. With permission.)

selection of both viroid-affected tissues and healthy control samples, due to the possibility that the viroid may not be uniformly distributed among all cells of a given host plant. In addition, temporal considerations may also influence the detection of quantitative differences between healthy and diseased tissues. Plasma membrane disorders similar to those cited for CEV-containing plants have been observed in virus-infected cells and found to decrease with age of infection.[47]

In ASV-infected avocado, both an increased frequency and altered internal organization of plasmalemmasomes was found in cells of strongly affected chlorotic areas.[42] This system provides the best definition of viroid-infected tissue and, therefore, minimizes some of the technical difficulties involved in defining and recognizing viroid-containing cells. Cultured tomato suspension cells derived from CEV-infected protoplasts or separated cells[20] also provide a more homogeneous sampling source in terms of viroid-infected cells. The increased frequency compared to healthy cells of paramural bodies or plasmalemmasomes in such CEV-containing cultured cells has also been reported. In the case of CEV infection, these structures accompanied the appearance of cell wall alterations, whereas in ASV infection, no changes in cell wall structure were reported. It, therefore, seems that the abnormal development of distorted or thickened cell walls cannot be directly correlated with aberrations in the paramural region. Nevertheless, these observations, when taken in perspective with the effect on other cellular structures, would seem to reinforce the view that stimulation of membrane pertubations occurs within the viroid-infected cell.

C. Chloroplast

Since chlorosis constitutes a major symptom expression of avocado sunblotch disease, as well as of CSV on mistletoe chrysanthemum,[39] it is not surprising that chloroplasts from viroid-infected tissue display disorganized structural features and inclusions. Contributing to the interpretation of this cytopathic effect as a direct result of viroid infection is the report that chloroplast disintegration was observed even in the absence of external symptom development with HSV or CPFV on tomato (*Lycopersicon esculentum* cv. Rentita).[19] The frequency of this reponse was far less than in conjunction with the more severe disease reactions as typical with CEV or PSTV on tomato. In these systems, in which gross morphological responses of the host to viroid infection are evident, widespread occurrence of chloroplast swelling was noted, followed by distortion of the thylakoid membrane system into an irregular arrangement of the grana and intergrana lamellae within the stroma. Breakdown of the outer chloroplast membrane and release of chloroplast contents into the cytoplasm eventually ensued (Figure 5). The apparent suddenness of the response in cells adjacent to seemingly unaffected cells and the occurrence in cells of chloroplast disruption concomitant with the absence of vacuoles suggests that the primary effect may be upon the tonoplast. The resultant release of hydrolytic vacuolar contents could initiate degradation of the outer chloroplast membrane. Although no differences in the size and number of plastoglobules (osmiophilic globules) was noted in this study nor in that of Hari[46] on PSTV-infected tomato, others have reported the occurrence of electron dense inclusions in chloroplasts of viroid infected tissues.[37,39,44]

In the infection of avocado with ASV, a similar initial response of chloroplast swelling was observed.[42] However, serial sections through the transition zone between the green and chlorotic tissue demonstrated a proliferation in the outer chloroplast membrane followed by dissociation of the internal lamellar system. Formation of crystalline bodies, which may result from coalescence of grana protein, occurred while the outer chloroplast membrane remained intact (Figure 6).[167] Therefore, it would appear that two distinct cytopathic effects are reflected on chloroplast structure, one involving degradation induced from without the organelle and a second affecting the internal structure of the grana. Further, it is possible that the cytopathic response is characterized by both a degradative reaction on mature chloroplasts and an interference in plastid development.

In constructing this comparative evaluation of data, we believe an awareness should be retained of the distinctions that exist, not only among the various host-viroid combinations, but also between the major classes of viroids. Of importance in this case is the fact that ASV has a more limited genome than PSTV or CEV, which is reflected in its highly restricted host range. It is not surprising, therefore, that the cytopathic effects of this viroid might also display characteristic distinctions from those of other viroids. The specificity of the reactions involved in the interaction of viroid with cell membranes, such as those of the chloroplast in this case, may reflect an intrinsic property of the particular viroid-RNA in question.

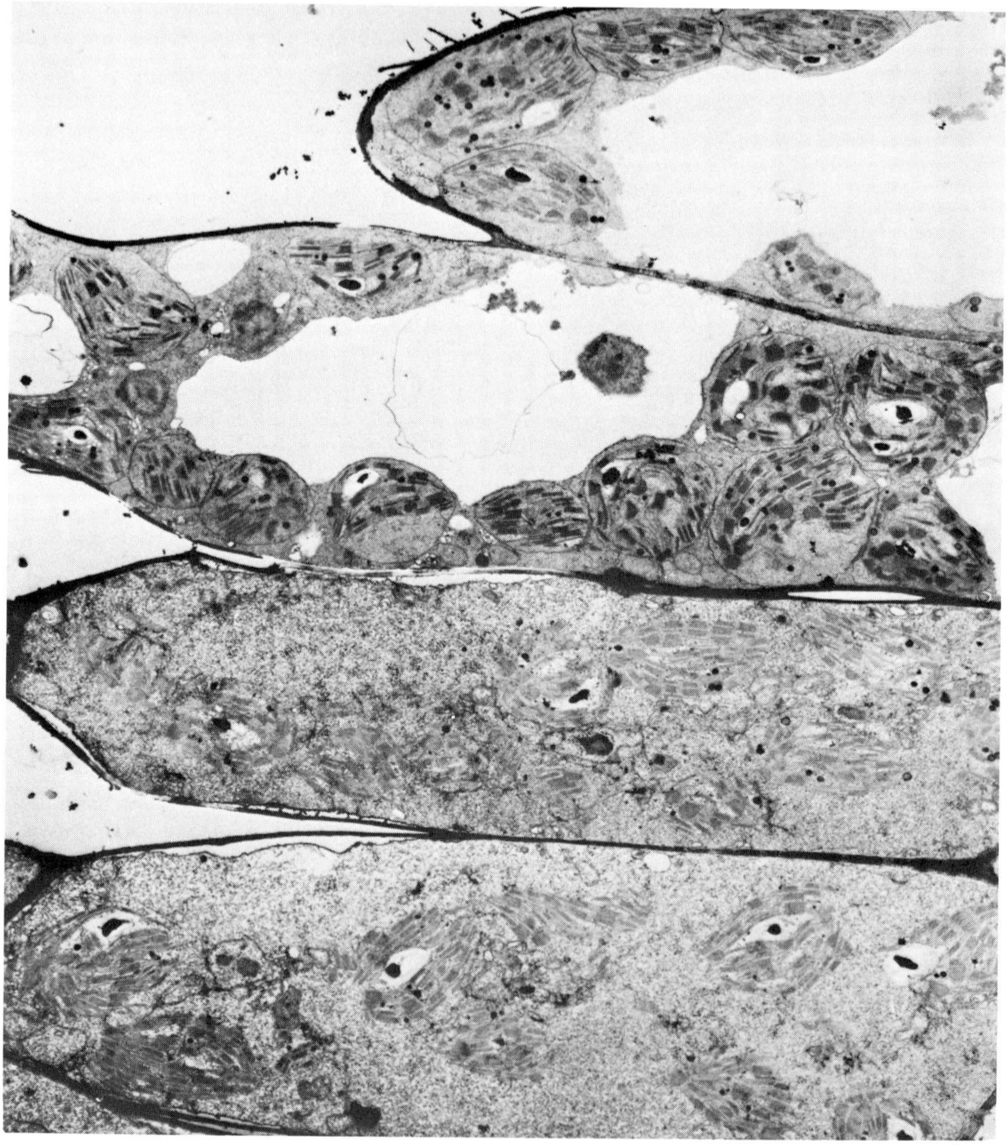

FIGURE 5. Cytopathic changes in the paliside parenchymatous cells of a citrus exocortis viroid-infected tomato plant. Note cells at top in which cell organelles are normally arranged and two neighboring affected cells showing disappearance of tonoplast and breakdown of chloroplast envelope. (Magnification × 5500.) (From Kojima, M., Murai, M., and Shikata, E., *J. Fac. Agric. Hakkaido Univ.*, 61, 219, 1983. With permission.)

It is very easy to dismiss the described chloroplast aberrations as simply the cellular expression of chlorosis and, therefore, probably a distantly related consequence in the pathogenesis cascade effect on any number of host processes. Certainly the particular effects are not viroid specific, for similar chloroplast features have been reported to accompany infection by a number of plant viruses and stress conditions. However, in the case of turnip yellow mosaic virus infection of chinese cabbage, replication of the virus was also found in association with chloroplast structures.[48] Also, even though viroids for the most part have

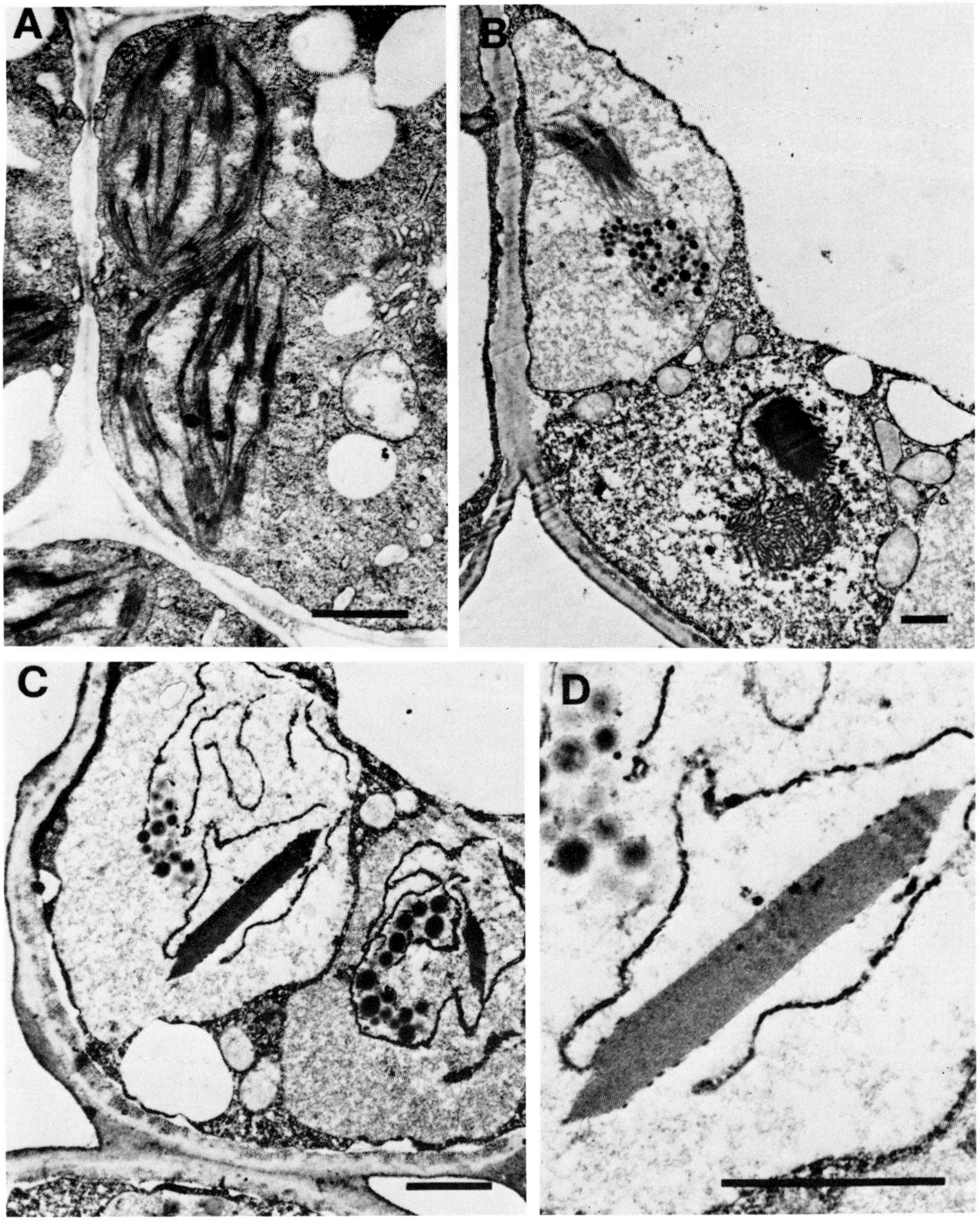

FIGURE 6. Chloroplast abnormalities caused by avocado sunblotch viroid (B) as compared with healthy chloroplast (A) in avocado leaf cells. Disappearance of grana structure (C) and formation of crystalline bodies (D). Bar represents 10 μm. (Unpublished data courtesy of Dr. P. Desjardins and P. Sasaki.)

been implicated as nuclear agents whose replication is accomplished within this organelle, at least for ASV, significant concentrations of the viroid-RNA have been found in association with chloroplasts and endoplasmic reticulum.[36] Furthermore, the major accumulation[43] as well as synthesis[168] of ASV occurs in the chlorotic or strongly ''bleached'' areas of avocado leaves. A similar, yet unusual reaction was induced in PSTV-inoculated tomato grown under continuous light. The new growth of viroid-infected plants displayed albino symptoms and contained three- to tenfold greater concentrations of the viroid than mature green tissue as determined by infectivity.[49] This reaction was more pronounced at 30°, whereas infected plants held at 16° remained green, as did healthy plants under all treatments. Since this reaction could be produced only on leaves which had not completely expanded prior to placement under continuous light, the presence of the viroid, in this case, would appear to have elicited a marked effect on plastid development, rather than to have disrupted fully differentiated chloroplasts. The relationship between the induction of prolamellar bodies and increased viroid content remains to be determined.

However circumstantial these lines of evidence must be acknowledged to be in the absence of definitive data, the broad consensus for the occurrence of viroid disease affects on chloroplasts must be recognized. Furthermore, as pointed out by Momma and Takahashi,[37] "the abnormal development of thylakoid membranes of chloroplasts can be functionally related to inhibition of the chloroplast differentiation which partially requires the nucleus-mediated gene expression". Determination of whether an indirect relationship exists between the biological activity of viroid-RNA as an abnormal snRNA and host organelle development or a direct interaction of the viroid occurs with the cytoplasmic membrane presents a most intriguing prospectus for future investigation. The potential of the chloroplast as an autonomous genetic system containing DNA and RNA and protein-synthesizing capacities, as well as its capability for reproduction by division, must also be considered in this discussion of the interaction of this organelle with a replicating heterologous RNA with the unique features of the viroid-RNA.

D. Tonoplast

The disappearance of the tonoplast (Figure 5) has been reported in conjunction with the disintegration of chloroplasts in HSV and CPFV infection of cucumber and PSTV and CEV on tomato. This phenomenon occurs when gross symptoms are expressed, as well as in the symptomless infection of tomato by HSV and CPFV.[19] In fact, it might be suggested that the release of vacuolar contents resulting from tonoplast disturbance causes the generalized disruptive affect on the chloroplast outer membrane. Some measure of specificity for, or intrinsic sensitivity of the chloroplast membrane possibly exists, however, since mitochondria are not similarly affected. Other than the deposition of electron-dense materials on the vacuole membranes in PSTV infection of the local lesion host, *Scopolia,*[9] no other similar tonoplast responses have been reported in association with viroid disease.

E. Mitochondria

Viroid effects on mitochondria described in two reports which center on either cells adjoining necrotic cells[9] or the envelopment of these organelles by chloroplast membrane[39] must be viewed as secondary effects induced by a degenerating cell condition expressed in the hypersensitive response or as a result of the primary activity upon chloroplast membranes.

F. Summary

In summary, we find it most interesting that this consideration of cytopathic structures appears to emphasize the basic segregation that exists between viroid replication and viroid pathogenesis, in which the principle site of viroid synthesis, the nucleus, does not display any deleterious effects as a result of sustaining substantial levels of a foreign, biologically active RNA. Secondly, it becomes apparent that no unique, viroid-specified cytopathic lesion has been described. As has been indicated in several of these reports, the changes in cells in response to viroid have been described in several cases of plant virus infection, stress conditions, or even healthy plant development. It is significant, nevertheless, that in the absence of any effect on the nucleus, a number of the cell membrane systems appear to be affected. Since the membrane surface is the physical site of recognition reactions for a number of regulatory processes, we might speculate that the viroid itself or a viroid-induced "signal" may be capable of interacting with these reactive interfaces. Finally, since the ultimate size and shape of cells is determined by cell walls, it is not surprising that a plant responding to viroid infection with severely delimited growth and development should display cell wall abnormalities.

VI. VIROID DISTRIBUTION WITHIN THE CELL

One approach in attempting to correlate cytopathic responses of host tissues to the primary action of an infectious RNA focuses on analyzing for the presence of the pathogenic agent at the site of action. The studies made on various viroid-host combinations have employed detection methods involving either bioassay procedures or visualizing the distinct viroid band after electrophoresis of nucleic acid extracts in polyacrylamide gels. Even though we have made an attempt to rationalize these data in Table 3, we must caution that no directly comparative or substantiating experimentation has yet been made with any one system. Therefore, the variation in diseased tissues as well as in procedures used to isolate all host cell constituents and viroid-RNA must be recognized. Also, some studies have focused at the outset on the isolation and purification of a particular organelle, principally nuclei, and therefore a complete "balance-sheet" of distribution of the recovered viroid is difficult to determine.

Because of the variations in the presentation of data, the summary presented in Table 3 cannot be taken as a strict quantitative comparison among the different studies. In most cases, the procedures employed relied on the separations achieved by differential centrifugation to generate preparations enriched in specific cellular components. Even though some samples were banded in density gradients, the ultimate purity of all fractions can be questioned. This is especially true of the intermediate range fractions pelleted at 10,000 to 80,000 $\times$ *g* for 10 to 20 min, which can comprise a complex mixture of organelles and membrane systems. Therefore, the fractions described in table 3 should be viewed as indicating broad categories rather than purified cellular constituents.

The trend most apparent from these studies is the strong indication for the association of infectious viroid with the nuclei-rich fractions. In most cases, the nuclei-rich preparation constitutes the fraction containing the highest concentration of viroid. Only in the ASV-infected avocado[36] system is the nuclear sample relegated to a role of secondary importance. It is interesting to note that this is the only study which does not utilize a viroid represented in the principal PSTV class of severe stunting agents. The low level of detection of TPMV from the nuclear fraction of infected tomato stems[50] may be related to the choice of tissue. The remainder of the studies presented in Table 3 employed infected leaves as starting materials.

Early studies with PSTV-infected tomato[51] and CEV-infected *Gynura*[52] produced highly compatible data, indicating the almost exclusive association of infectious viroid-RNA with purified chromatin. However, in succeeding studies which employed various systems involving either the same extraction medium with HSV-infected hops or cucumber,[53] CEV-infected *Gynura* extracted in a different medium;[54] or a similar viroid, TPMV, in tomato, in yet a different medium[50] significant concentrations of viroid were detected especially in the postnuclear, organelle-membrane fractions.

Included in these studies have been attempts to define the extranuclear site of viroid association by monitoring a series of sequental differential centrifugation samples and relating the infectivity of each fraction to the total recovered. Three reports involving CEV in *Gynura,*[54] TPMV in tomato,[50] and HSV in hops or cucumber[53] implicating a membrane fraction have been presented. Density determinations of the complex membrane fraction indicate primarily a plasma membrane-like component. With ASV in avocado, the chloroplast

Table 3
SUBCELLULAR SITES OF VIROID ASSOCIATION

		Organelles				
Viroid	Host	Nucleus	Membrane	Ribosome	Cytoplasm	Ref.
ASV	Avocado	+	+ + + +	±	−	36
CEV	*Gynura*	+ + + +	−	−	−	52
	Gynura	+ +	+ + + +	±	−	54
	Gynura	+ + +		+ + +		55
		+ + + +		+		
HSV	Hops } Cucumber	+ +	+ + + +	+ +	+	53
PSTV	Tomato	+ + + +	±	−	−	51
	Tomato	+ + + +	±	[ ]	[ ]	26
TPMV	Tomato					
	Leaves	+ +	+ + +	±	−	50
	Stems	+	+ + + +	+	−	

or endoplasmic reticulum fraction contained by far the greatest concentration of viroid as detected by PAGE. Taken together, these data present a fairly compelling suggestion that the viroid-RNA is not rigidly restricted to the nucleus of infected plants.

Determination of the subcellular structural locus of a specific molecular moiety like viroid-RNA is fraught with technical problems including those associated with (1) disruption of cells, (2) separation of normal cell components complicated with the possible presence of unusual structures in the diseased condition, and (3) the unknown in vivo intermolecular association(s) of the viroid-RNA. Recognizing the primary importance of the nucleus in viroid synthesis and the porous nature of this structure raises the question as to whether we might expect the viroid-RNA to be exclusively retained therein during purification. If leakage does occur, what governs the various possible association(s) of viroid-RNA with extranuclear components? In light of such considerations, strict experimental controls are exceedingly important, yet themselves difficult to interpret.

When a tRNA preparation from CEV-infected *Gynura* was added to a crude extract from healthy *Gynura* which was then subjected to a separation treatment similar to that used for extracts of infected tissue, it was possible to recover infectious viroid from the postribosomal supernatant.[54] It was encouraging to note that the exogenously added purified RNA was not completely degraded during the experimental manipulation. However, substantial quantities of the added CEV were also recovered in the nuclear and membrane fractions. This result emphasizes the problem intrinsic in the disruption of tissue which produces the admixing of constituents previously separated by compartmentalization. Additional complications of interpretation are introduced by use of this control, since the viroid was exogenously added as a purified free-RNA and, as has previously been indicated, no compelling evidence has yet been presented to establish that the in vivo nature of the viroid is indeed as a free-RNA. Therefore, if the viroid exists as a viroid-RNP in the cell, this control might not be strictly appropriate.

Nevertheless, more recent studies have attempted to address this question of potential nonspecific association of viroid-RNA with postnuclear host constituents. By employing a low ionic strength buffer in a high medium/tissue ratio, which reduces leaching of nuclear enzymes during homogenization,[55] it was possible to prevent the adherence of exogenously applied CEV to pelleted subcellular host components.[169] When these conditions were applied to extractions of CEV-infected *Gynura,* 60 to 80% of the viroid infectivity was associated with a 1000 g $\times$ 10 min nuclei-rich pellet. However, significant quantities of viroid remained in the supernatant containing organelles and membranes. Furthermore, it is likely that even the 1000 g pellet might have contained some of the larger sheets of dense membranes produced during homogenization.

A recent report by Schumaker et al.[26] focused on the localization of PSTV in subnuclear fractions from infected tomato leaves. It was estimated from the appearance of viroid bands after PAGE of pelleted nucleoli that about 95% of the cellular PSTV was associated with this nuclear component. Once again, the problems resident to the disruption of intact cells are reflected in attempts to fractionate nuclei. The localization of nuclear associations of viroid are further complicated by the absence of membrane barriers to separate nuclear organelles or to define compartments within matrixes as reactive as the chromatin and the nucleolus. The release of PSTV from the nucleolar pellet by high salt treatment (500 m*M* NaCl, 50 m*M* Mg^{2+}, 10 m*M* Tris HCl) was taken as evidence for the existence of the viroid-RNA in the nucleolus as a protein-RNA complex. Since earlier studies by Diener[51] and Sänger[52] had demonstrated the release of viroid from purified chromatin with a tenfold lower concentration of phosphate buffer, in addition to such treatments as freezing and thawing and exposure to 1% SDS, 6 *M* urea, or 2 *M* guanidine-HCl, it would seem presumptuous to imply that the high-salt data confirms a protein-viroid-RNA association, even though the existence of a viroid-RNP is conceptually a most attractive proposition.

The suggested association of the viroid with the nucleolus permits us to pose an interesting question. Since complete nucleolar dissociation and reassociation occurs during mitosis, is the viroid-RNA or the putative viroid-RNP governed during that process by the same host control process, thus permitting the containment of the suggested concentration of greater than 95% of cellular viroid within the nucleolus? The stable titer of viroid detected in repeatedly transferred cell suspension cultures suggests a most highly integrated host control function, if almost exclusive association of viroid with the nucleolus does in fact occur.

It would seem that the most cautious conclusion to draw from these data would be to accept the compelling experimental evidence for association of the viroid with the cell nucleus. Such a conclusion seems particularly justified, based on the demonstration of viroid synthesis in cell-free, nuclei-rich preparations.[56,57] At the same time, it remains highly probable that some significant or detectable portion of the viroid content of the cell can be found in association with some yet ill-defined extranuclear membrane component. Even Schumaker et al.[26] recognized that " . . . one cannot conclude that the cytoplasm or other cell organelles are completely devoid of viroids".

It must be kept in mind that in all of these studies, analysis of viroid presence has been based either on recognition of a defined RNA band after PAGE or the ability of infectious components of extracts to cause disease. In either case, we might question whether the molecular species identified by the diagnostic procedure is necessarily the same as the intracellular molecule presumed to possess biological activity. Therefore, with this logic, a seemingly minor, well-placed fraction of a total population might remain exceedingly important in the process of pathogenesis. And as a corollary, subnuclear associations of especially high concentrations of viroid-RNA may relate to viroid processing events such as the formation of circular forms from infectious linear molecules by a host ligating or splicing mechanism as in the processing of pre-mRNA or pre-rRNA. Finally, we have focused on the viroid molecule and should be receptive to or at least congnizant of the viroid complementary molecules that occur in the cell. Unfortunately, no data is yet available on the cellular distribution of this viroid-related molecule.

VII. CULTURE SYSTEMS FOR VIROID-CONTAINING CELLS

With the dramatic effects on plant development and cell structures elaborated in the process of viroid disease expression, as well as the rather specific association of viroid-RNA with subcellular constituents, it might be anticipated that significant metabolic aberrations may be detected in viroid-containing tissues. For rather obvious reasons, during attempts to address this issue, attention has been focused on the direct involvement of viroid-RNA on nucleic acid and protein metabolism. In addition, secondary effects on hormonal control processes and carbohydrate constituents of the cell wall structures which may provide leads to the gross symptom expression of stunting have been examined. However, before we introduce and attempt to correlate data contributed in these diverse fields, it will be necessary to first describe some of the alternative systems to whole plant or tissue fragments which have been exploited in many of these studies.

In 1974, Zaitlin and Beachy[58] predicted that " . . . many fundamental questions as to how viruses replicate, and the biochemical effects they induce in their hosts will be investigated . . . " utilizing separated cells, protoplasts, and small callus pieces. It was suggested that these systems could provide a vehicle whereby primary metabolic lesions introduced by the pathogenic nucleic acid may be exposed in advance of an obvious "sick" condition and the concomitant secondary cascading effects of virus infection. Much was anticipated from systems that permitted the maintainance of relatively homogenous, infected tissues in controlled environments.

Early events in the initiation of virus infection have been analyzed principally with protoplast and separated cell systems. These studies are made possible by the inoculation of cells derived from healthy plants with relatively high concentrations of virus and are usually short-lived experiments (0 to 72 hr). Such experiments are directed at detecting intermediary molecules in virus replication and are not amenable to studies of pathogenesis. Although callus tissue cultures provide a system for studies of longer duration, the problems involved in inoculation and maintainence of homogenous, virus-infected cultures have discouraged their use in biochemical studies. Although evidence for persistance of plant viruses in callus cultures has been presented in the review by Raychaudhuri,[59] there persists a widespread perception that spread and retention of the viral nucleic acid concentration in callus systems remains a problem limiting its application in such systems.[60]

Viroids, because of the RNA nature of the infecting molecule and the low concentration of progeny replicated in infected cells, present additional technical problems, especially when investigating early events in the initiation of infection.[14,15] However, the unusual property of an intimate association of viroid with the nucleus as the locus of both replication and accumulation potentially provides a marked advantage in maintaining stable cultures of infected cells. That is, the persistence and stable titer of viroid-RNA in cultured cells may be unlike that experienced in any virus-infected cultures. With this condition it becomes feasible that metabolic lesions involved in the expression of viroid pathogenesis may become clearer and even more apparent in cultured cells than in whole plant systems. An observation introducing this possibility was made when tomato tissues were doubly infected with the citrus exocortis viroid and the tumor-inducing bacterium, *Agrobacterium tumefaciens*.[13] When tumors were produced on stems of CEV-infected plants, the viroid-RNA as well as viroid complementary sequences were detected within the tumors in concentrations far greater than normally present in viroid-infected stems or apex tissues. This association of viroid with mitotically active, heterogeneous outgrowths of intact plants stimulated interest in the development of callus and suspension cell cultures.

Callus and cell suspension cultures were derived from PSTV-infected tomato, *Lycopersicon esculentum*[61] and *Lycopersicon peruvianum*,[62] and potato, *Solanum demissum*.[63] Callus produced from stem segments retained concentrations of PSTV in excess of that found in leaf tissues. In newly established potato cell suspensions, a value of 10% of total cellular RNA synthesis was achieved for PSTV synthesis, whereas synthesis of the viroid in tomato protoplasts, cell suspensions, and green tissue reached only 0.1 to 3%.[63] For the most part, these studies were directed to establishing a credible system of long-term replication for viroids, identifying complementary molecules involved in viroid synthesis, and defining conditions of optimum growth of both cells and viroid contained therein. The problems of low viroid concentration and synthesis resulting in extremely poor detection of the viroid-RNA encountered in protoplast systems were thus overcome in the callus and suspension cultures. Even though Zelcer et al.[61] reported a reduced synthesis of viroid in one cell line with time, it was generally accepted that viroids in cultured cells ''behave differently from conventional plant viruses''. However, since inoculation of cultured cells was not possible, the callus systems are more amenable to studies of established continuous viroid replication rather than early events in the initiation of infection.

Reports of both no differences[62] in growth characteristics or morphological appearance between infected and noninfected cultures, as well as a slower growth rate of cell lines infected with a severe isolate of PSTV[61] have been made. The suggested absence of either a qualitative or quantitative distinction in the protein profile from healthy and infected cell lines[61] is difficult to verify from the complex protein patterns provided. The comparative data provided in these reports might be affected by the heterogeneous origin of the cell lines which, although sustaining viroid replication, appear to display changes in properties with continuous subculturing.[170] No attempt was made in these studies to determine if infected cell lines were totally comprised of viroid-containing cells.

Utilizing a hybrid tomato (*Lycopersicon esculentum* × *Lycopersicon peruvianum*) infected with CEV, Marton et al.[20] introduced an alternate procedure to produce cell lines of more homogeneously infected cells. Either protoplasts isolated from infected leaf tissues or separated cells derived from leaf callus and suspension cultures were plated onto a feeder layer culture system as previously employed to recover single cell clones.[64,65] In the absence of nurse cells in the feeder layer no colonies developed. When 71 separate colonies were recovered, increased, and subsequently analyzed for viroid-RNA, 100% contained CEV. In contrast, 60% of cultures derived from TMV-infected tissues proved to be composed of healthy cells.[66] This unexpected result seemed to indicate that this method of preparation and cell plating in some yet unknown way constituted a selection procedure for viroid-infected cells or that, in fact, 100% of the original cell and protoplast populations carried the viroid.

Although this technique does not permit the description of these cultures as clones derived from single cells, it would seem probable that they did originate from either single cells or a few cells from a single cell source. Lines of these CEV-infected cells subcultured as solid callus over a period of about 4 years, to date, have not changed in characteristics, including viroid titer. In fact, the enhanced performance of CEV-containing cells was observed when callus cultures were held beyond the usual 4-week subculturing period.[38] During the usual interval between subculture, calli derived from healthy and CEV-infected cultures were indistinguishable in color, texture, and friability. However, when healthy calli were maintained on the same medium for longer than 6 weeks the cells became dark (Figure 7). New cultures initiated from callus from such plates displayed a diminished growth rate. CEV-infected cultures aged for a similar period maintained normal growth characteristics and were subcultured without any noticeable effect.

When the growth rates of calli of cell lines developed from healthy tissue by the cell plating procedure were compared within a 4-week subculture period to the rates of growth of CEV-containing lines, no dramatic difference could be detected.[20] Therefore, under certain controlled conditions of culture, the viroid-containing cells could be thought of as asymptomatic with respect to both morphology and growth rate. Another, even more surprising aspect was observed when cultures were subjected to increasing temperatures. The infected cells performed significantly better in terms of growth rate and cell mass at 30°C than cells derived from healthy tissues. Whether this enhanced growth at elevated temperatures of callus cells containing viroid-RNA is related to the enhanced viroid replication at 35°C in

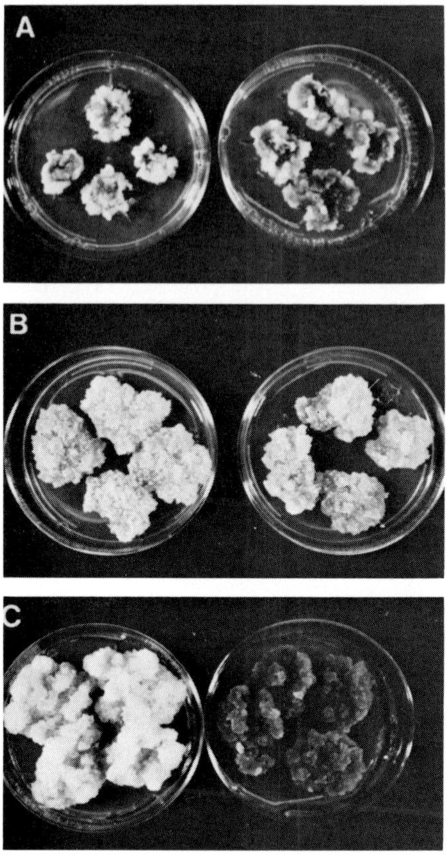

FIGURE 7. Healthy (right) and citrus exocortis viroid (CEV)-infected (left) callus cultures: (A) initiation of callus from leaf explants (4 weeks in culture), (B) callus cultures after 20 serial subcultures (4 weeks in culture since last transfer), and (C) callus cultures after 20 serial subcultures (8 weeks in culture since last transfer).

intact plants[67] is only conjecture at present. In fact, enhanced synthesis of viroid-RNA at 35°C over 25° was reported for CPFV-inoculated protoplasts,[15] even though the reduced incorporation of ^{3}H-uridine into host 5S and 4S RNA at 35°C would have suggested a deteriorating system. Another, perhaps more remote, possibility is that the viroid may stimulate the production of specific host proteins that enhance the survival of cells under normally lethal temperatures, as in the heat shock reaction observed as a general phenomena in animal and plant cells.[68,69] Nevertheless, it appears the environmental conditions such as temperature (in this case) may be exploited to discriminate between cultured healthy and viroid-containing cells in a common medium. The effects registered on cell growth in response to changes in the medium constituents such as carbon source and phytohormones will be presented later.

These cultures represent a good model system to investigate fundamental metabolic lesions introduced by viroid infection. This is not to say that many of the effects elaborated as a result of viroid infection might not require the cell-cell interactions that function in the normal progression of tissue differentiation and development characteristic of intact plant tissues. Additional properties of viroid-containing cell cultures will be introduced in the section addressing specific processes involved in hormonal and carbohydrate metabolism. Whether these events represent primary lesions in viroid pathogenesis remains a leading question. Continuously subcultured callus cells of plants undergo considerable alteration in basic genetic characteristics. Therefore, the danger of selecting for a resultant aberration in a host-specified trait in the absence of any action by a foreign molecule such as the viroid always remains a possibility. The most convincing evidence that a property of a cell line is derived from viroid infection would be furnished by demonstrating either the loss of that particular property in cultures cured of the viroid or the introduction of that trait upon infection of a healthy cell line with the viroid. Since neither of these procedures has yet been reported for viroid-cell line interactions, we are left with the task of simply presenting the available experimental observations.

VIII. METABOLIC PROCESSES IN VIROID-CONTAINING CELLS

When attempting to decipher primary metabolic events which result from a pathogenic RNA in relation to the characteristic stunting symptoms of viroid-infected tissues, our attention is naturally directed to (1) nucleic acid and protein processing because of the RNA composition of the infecting agent, (2) hormonal control because of the predominance of the altered growth response involved in the stunting reaction, and (3) carbohydrate composition because of the principal cytopathic responses of cell walls and plasma membranes. As is true of most instances when attempts are made to define the mode of action of pathogenesis caused by fungal and bacterial agents and even by plant viruses on plant systems, the information or observations are usually spotty, at best. Unlike physical characterization or even the elucidation of mechanisms of replication, a molecular description of pathogenesis, in most cases, encompasses much conjecture supported by a patchwork of observations. Nevertheless, the ultimate simplicity of a pathogenic molecule as presented to us in the viroid with its mere 250 to 350 nucleotides provides sufficient incentive to encourage this undertaking with even bold strokes.

A. Nucleic Acid Synthesis and Viroid Infection

Studies concerned with nucleic acid synthesis in viroid-infected cells have largely focused on demonstrating the replication of the viroid-RNA. Utilizing a system of ^{32}P root-feeding of PSTV-infected tomato, Hadidi and Diener[70] reported that labeled viroid-RNA represented only about 0.08% of the total RNA and only about 0.4% of the 2 *M* LiCl soluble low molecular weight RNA. The rate of incorporation into PSTV was not affected by increasing times of incorporation up to 5 days, which was in contrast to the decline in uptake after a similar time noted with a conventional plant virus infection such as BMV.[71] The possibility of a greater coordination of viroid synthesis with host nucleic acid synthesis was thus suggested.

In a direct comparison of the incorporation of ^{14}C-uracil into the nucleic acids of healthy and PSTV-infected rapidly growing potato sprouts, it was observed that the rate of incorporation into diseased tissue was consistently greater.[72] The ratio of incorporation into healthy vs. infected tissues increased from 0.2 at 16 hr to about 0.5 at 46 hr. It would appear, therefore, that the differences in incorporation rates between healthy and infected tissues may be due to a differential uptake of the isotope, with the healthy tissues slower to achieve pool saturation. This suggestion is supported by the data presented for incorporation into the 4S and 5S RNAs and DNA which indicate a consistent percent of total incorporation. An unusual feature of this data is that the percent incorporation into the 9S (reported as 7S by others) RNA in the three feeding times reported was consistently lower in diseased tissues. In fact, in all cases, the percent incorporation into the 9S RNA of healthy tissues was about equal to that of the 9S plus PSTV in the infected tissue.

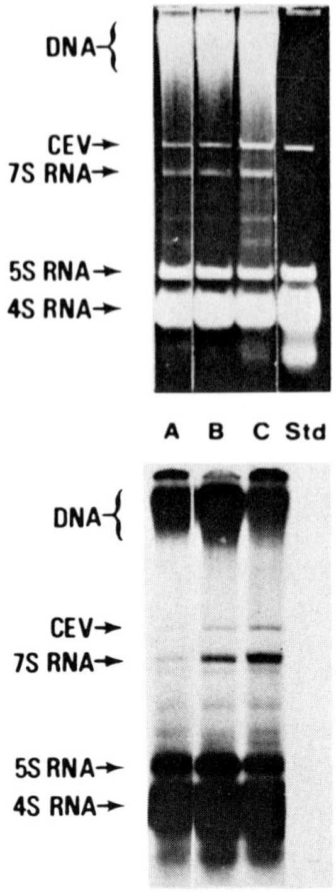

FIGURE 8. Polyacrylamide slab gel (5%) electrophoresis pattern of the 2 *M* LiCl soluble nucleic acid fraction obtained after 1 hr ^{32}P incorporation into tomato suspension cells containing citrus exocortis viroid (CEV). (Top) Ethidium bromide stained patterns of nucleic acid extracted from equal weights of suspension cells at (A) exponential, (B) linear, or (C) stationary stage of growth, compared to a CEV-containing nucleic acid preparation from *Gynura aurantiaca* (standard). (Bottom). Autoradiography of the same gel after 3-day exposure to X-ray film.

This indication of a similarity in the rate of synthesis of a specific host RNA and the viroid was not supported by studies of PSTV in tomato cell suspensions[61] when viroid synthesis was correlated with cell growth stages. Since viroid synthesis appeared relatively higher during the log and stationary phases of culture, it was interpreted that viroid replication was not closely linked with the synthesis of other host RNA species.

However, when Lin[41] monitored the rate of synthesis of specific nucleic acid species during the cell growth cycle (Figure 8), a close coordination of viroid synthesis with 7S RNA synthesis was observed. When a comparison was made of the relative incorporation of a 1 hr ^{32}P pulse, the synthesis of viroid-RNA as compared to DNA or 5S RNA increased, whereas the ratio to 7S RNA remained relatively constant (Table 4). Not only was the rate of CEV synthesis maintained in stationary phase cells when compared to a host species like 5S RNA, but the rate of accumulation exceeded the constant level expected. Therefore, although a similarity exists between the rate of synthesis of the viroid RNA and a host species (7S), the viroid also displays a unique characteristic of higher levels of accumulation, perhaps reflecting its totally unique and highly stable structure. Whether the like kinetic relationship between 7S RNA and the viroid reflects a functional comparability or simply a fortuitous result of similar regulation is unclear.

When cells at distinct growth stages were subjected to a continuous source of ^{32}P for 48 hr, the incorporation rate into viroid-RNA closely paralleled that of the DNA and the host small RNA population.[41] This would indicate that the foreign RNA is synthesized and regulated in reasonable harmony with the host nucleic acid and might be viewed as assuming

Table 4
RELATIONSHIP AMONG NUCLEIC ACID SPECIES SYNTHESIZED IN CEV-INFECTED CELLS AT DIFFERENT STAGES OF THE GROWTH CYCLE

Growth stage	CEV/DNA[a]	CEV/5S RNA	CEV/7S RNA
Exponential	1.05	1.08	26.1
Linear	1.52	2.0	19.7
Stationary	8.56	5.27	18.3

[a] As determined from pulse labeling (1 hr) with ^{32}P followed by phenol extraction. All species contained in 2 *M* LiCl soluble fraction analyzed by PAGE and autoradiography.

the role of a cellular small nuclear RNA. This perspective is supported by the stable titer and persistence of the viroid-RNA in cell culture systems.

B. Protein Synthesis and Accumulation

With the definition of the infectious viroid as a free-RNA molecule of monocistronic dimension (10^5 daltons), theoretically capable of serving as a template for a protein of about 10,000 daltons, studies were initiated in the mid-1970s to determine if in vitro translation was possible. Both PSTV[73] and CEV[74] were tested in several protein synthesizing systems including wheat embryo *(Triticum aestivum)* and *Escherichia coli*. In addition, a system using *Bacillus stearothermophilus* at 60° was employed to permit partial denaturation of the secondary structures of the viroid-RNA. The low level of amino acid incorporation (0 to 4.7%) was comparable to the stimulation observed when nonmessenger RNA such as the 16S ribosomal RNA from *E. coli* was added to the reaction. Electrophoretic analysis of the small amount of isotope incorporated did not reveal a discrete product.

In the CEV study by Hall et al.,[74] a further test was conducted to determine if the viroid-RNA might interfere with the synthesis of a known messenger, bromegrass mosaic virus(BMV)-RNA_4. The addition of CEV to the synthesizing system had no effect on the quantitative level of incorporation nor the quality of the product produced. As an additional experimental approach to this question, suggested by the ability of aminoacyl synthetase enzymes to catalyze the binding of amino acids to viral RNA, the aminoacylation of CEV was attempted and found not to be possible. When freshly taken oocytes from *Xenopus laevis* were injected with CEV under conditions in which BMV-RNA and hemoglobin mRNA were translated, no viroid-specified protein could be detected.[75] Therefore, the viroid-RNA appeared to be totally inert in these systems and it would seem reasonable to conclude that viroid pathogenesis is not mediated by a viroid-specified protein.

A suggestion was made on theoretical grounds by Matthews[76] that the viroid might constitute a negative-strand virus whose complementary RNA[77] could function as a messenger for one or two viroid proteins. However, when the nucleotide sequences of PSTV and CSV, which display about 70% homology, were analyzed for theoretically possible translation products from either of the viroids or their complementary sequences, major differences between the putative polypeptide translation products of each suggested that common proteins were not coded by either strand of both viroids.[78]

When PSTV-infected tomato leaf tissues were compared with healthy leaves, the ratio of leucine-^3H incorporation suggested that no viroid-related, low molecular weight protein was produced.[79] The indication that synthesis of two high-molecular weight (155 and 195 $\times$ 10^3 daltons) proteins similar to those detected in TMV infection were enhanced in infected tissues was not clearly defined and has not been further clarified to date. More recently, results of protein synthesis have been obtained from suspension cells derived from healthy and PSTV-infected tomatoes.[61] Proteins synthesized in early log phase cells and analyzed by two-dimensional electrophoresis displayed neither quantitative nor qualitative changes resulting from the presence of the viroid.

Primary infection of *Gynura aurantiaca* by CEV often results in an interesting pattern of symptom expression in the first leaf affected by the viroid. It is possible to recover from the same leaf an apparently unaffected half-leaf, as judged from the gross symptoms and absence of detectable viroid, and a CEV-infected half-leaf.[11] This effect has already been described as exploited for controlled investigation of cytopathic effects. When protein extracts as well as lysed protoplasts from these tissues were compared, a quantitative increase in host-specified proteins of about 15,000 daltons (CEV-P_1) and 18,000 daltons (CEV-P_2) was detected in samples derived from infected half-leaf tissue.[10] Since preparation of the protein samples included boiling in excess SDS, the CEV-P_1 and CEV-P_2 may represent low molecular weight polypeptides or subunits of one or more complex proteins. The viroid-enhanced proteins could be recovered from the postribosomal, high-speed supernatant fraction from infected tissue and increased as a proportion of the total protein extract with prolonged infection.

Confirmation for the host-specified nature of viroid-related protein was made by Conejero et al.[80] wherein CEV infected *Gynura,* tomato cv. 'Rutgers', potato cv. 'Marijke', and citron were reported to have displayed intensification of low molecular weight proteins. However, the size of these proteins was dependent upon the host. The estimate for CEV-P_1 varied from 13,700 (corrected from the previously reported 15,000) to 12,000 daltons, depending upon whether infected *Gynura* and citron or tomato and potato were analyzed, respectively. Similarly, the CEV-P_2 varied from 18,000 to 16,300 daltons, as influenced by the same host relationship. In uninfected *Gynura* plants maintained for 3 years, proteins identical in size to CEV-P_1 and -P_2, termed senescence proteins (Sen-P_1 and -P_2) were detected as apparent products of natural senescence (Figure 9).

The enhancement of production in aging healthy plants of the P_1 and P_2 polypeptides indicates that neither is a translation product of the viroid-RNA or any mRNA exclusively related to host-viroid interaction. This experimental evidence, taken with the previous theoretical consideration of possible translation products from PSTV and CSV,[78] does not support the proposal by Matthews[76] for polypeptides specified by CEV-complementary sequences, although such complementary molecules have never been isolated in sufficient quantities to be tested in in vitro translation systems. Thus, the possibility exists that common events are involved in viroid pathogenesis and the natural senescence process in healthy plants.

When protein extracts from CEV-infected *Gynura aurantiaca* were fractionated into acid (0.25 N HCl) soluble and insoluble proteins, the bulk of P_1 was precipitated by the acid treatment, although traces of protein banding at the P_1 position were observed from the acid-soluble fraction. Protein banding at the P_2 position was detected in both acid-soluble and -insoluble partitions.[10] This would indicate that the proteins detected after SDS-PAGE may comprise a complex of similar sized, yet distinct proteins. The acidic nature of the principal component of the *Gynura* P_1 protein population was confirmed by Flores et al.[81] who determined the isoelectric point of 5.21 for purified preparations.

More recently, the scope of host-specified proteins which undergo enhanced production in the presence of pathogen(s) has been expanded to include proteins stimulated by infection of tomato by viroids (CEV, PSTV, CPFV, CSV) plant viruses (TMV, CMV) and the fungus, *Cladosporium fulvum.*[82] One of these proteins, the most dramatically enhanced in tomato plants, p^{14}, which probably coincides with the previously reported tomato CEV-P_1,[80] has a molecular weight of 14,000 daltons and an isoelectric point of 10.7.[83] This finding suggests

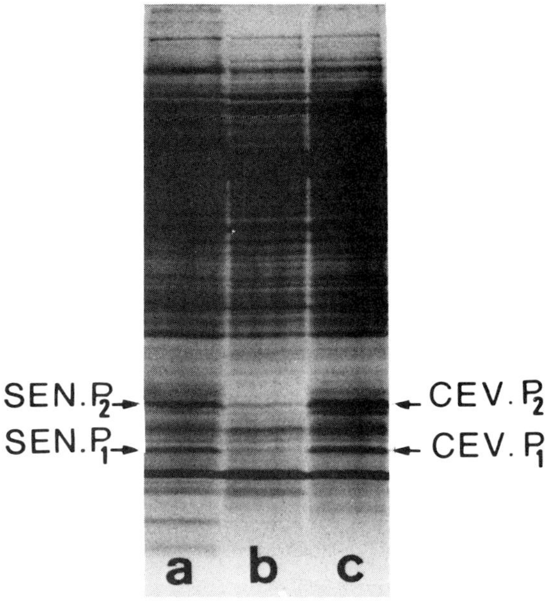

FIGURE 9. SDS-polyacrylamide slab gel (14%) electrophoresis of *Gynura aurantiaca* proteins from (a) 4-year-old healthy plants, (b) uninoculated controls, and (c) citrus exocortis viroid (CEV)-infected plants 1.5 months postinoculation of rooted cutting. Positions of CEV-P_1 and -P_2 and senescence proteins SEN-P_1 and -P_2 are indicated.

that the host dependency governing the nature of the enhanced proteins also involves properties other than molecular size. In addition, the decrease in a number of protein species in the range of 14,000 to 45,000 daltons has been reported in different hosts infected by CSV, posing the possibility that enhanced low molecular weight species may represent subunits of larger complex proteins. When tomato proteins were isolated employing an acid extraction procedure, a series of nine proteins (10,000 to 38,000 daltons) displayed an increased concentration, while four species (14,500 to 105,000 daltons) decreased as a result of infection by viroids, viruses, or *Cladosporium fulvum*.[84] All appeared to be host-specified and of lesser importance than the p^{14} species.[83] These observations strengthen the proposition that the enhanced occurrence of certain proteins which accompanies viroid infection is a host response which may be generalized beyond this particular class of host-pathogen interactions. The P_1-like (p^{14}) proteins may then be thought of as representative of the general class of ''pathogenesis-related'' or ''PR'' proteins,[85,86] according to the expanded meaning given to this term at the workshop on pathogenesis-related proteins in plants.[87] Within this definition were included all proteins in general, which are stimulated in plants as a consequence of the pathologically altered metabolism.[88]

With this expanded definition and the application of the pH 4.3 electrophoretic system of Reisfeld et al.,[89] which permits resolution of proteins of a more basic nature than the pH 8.8 system of Davis,[90] it has been possible to consider and detect new PR proteins. Figure 10 presents ten cationic proteins stimulated in CEV-infected tomato.[171] The C_2 species resolved in these native gels, which is the most dramatically enhanced protein, is a monomer with a molecular weight of about 14,000 daltons as determined by the method of Ferguson[91] and an isoelectric point slightly greater than 10. In SDS gels, C_2 comigrates with tomato P_1. Taken together, these data strongly suggest that C_2, tomato CEV-P_1, and tomato p^{14} are the same protein.

In contrast, only four cationic PR proteins were detected following acid-extraction of an alternative CEV-infected host, *Gynura aurantiaca,* again re-emphasizing the specificity of

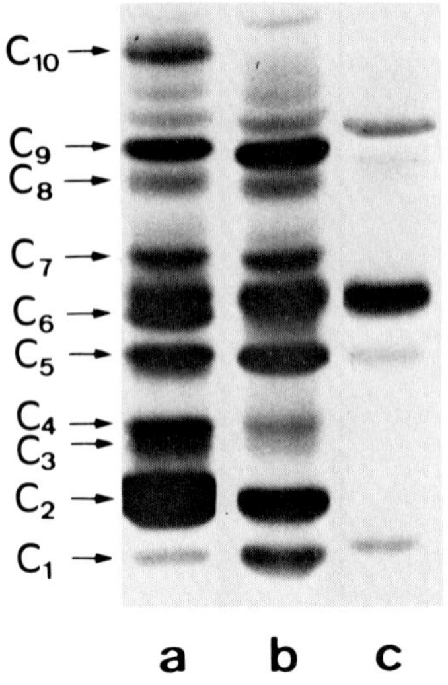

FIGURE 10. Acid (pH 2.8)-extractable native PR cationic proteins analyzed on polyacrylamide gels at pH 4.3[89] from tomato leaves either (a) infected with citrus exocortis viroid, (b) half-leaf pricked with 0.2 *M* ethephon, or (c) untreated healthy half-leaf.

host control in the biosynthesis of these proteins. The absence of new mRNAs as precursor to PR protein accumulation[92,93] indicates the constitutive nature of the PR-mRNAs, which is also compatible with the lack of mRNA activity of viroid-RNA.

Considering the rather dramatic responses elicited on host plant development by viroid infection, it should not be surprising that a more complex pattern than found in the host protein profile of the healthy plant should emerge. The major question with which we are left and to which very limited efforts have been directed is what is the biological activity or functional significance of these protein alterations?

One report of the purification of a protein enhanced as a result of viroid infection, the P_1 protein from CEV-infected *Gynura aurantiaca,* including an attempt to define its biological activity, has been made.[81] Preparations of the protein purified to include only one additional contaminating species of low concentration permitted definition of P_1 as an acidic protein (pI = 5.2) with an apparent molecular weight of 1.0 to 1.5 $\times$ 10^4 daltons as determined by gel filtration under nondenaturing conditions. This estimate is in agreement with those previously made for SDS-treated preparations and suggests the absence of a subunit role for CEV-P_1.

Ribonuclease activity was considered as a potential function for the CEV-P_1, since the molecular weight is about identical to pancreatic RNAse and increased ribonuclease activities are detected in diseased tissue. Although some preparations of P_1 did inhibit CEV infection when mixed with inoculum, as well as degrade other natural and synthetic RNAs, inconsistent results were reported.[81] Since symptom expression in response to viroid infection suggests a potential hormonal dysfunction, experiments to determine the binding activity of CEV-P_1 to the phytohormones gibberellin A_1/A_3 and β-indolacetic acid were also tested. No affinity was detected under either in vivo or in vitro conditions.

Therefore, although the enhanced production of discrete low molecular weight host proteins isolated from viroid-diseased tissues displaying an active metabolite turnover rate have been noted and some physical characteristics have been determined,[10] virtually nothing is yet known of the biological activity or cellular function of these moieties.

From the foregoing experimental as well as theoretical considerations, it becomes evident that protein alterations during viroid disease expression are derived from host-specified

FIGURE 11. *Gynura aurantiaca* DC showing stunting and leaf epinasty and rugosity produced by citrus exocortis viroid (CEV) infection or a high dosage (500 mg/ℓ) Ag^+ applied as aqueous foliar spray of $AgNO_3$ with 0.1% Tween® 80 as a wetting agent. To synchronize the evolution of symptoms, Ag^+ treatment was initiated at the onset of CEV-induced symptoms. Treated plants are compared to a healthy (H) control.

components. Therefore, an alternate approach to the questions of functional significance can be directed to simulating in uninfected hosts the effects produced in viroid-infected plants and, by analogy, proposing a logical mode of action.

The characteristic symptoms of stunting and leaf epinasty of viroid diseases resemble the effects of ethylene on healthy plants. Spraying of tomato plants with ethephon (2-chloroethylphosphonic acid), which liberates ethylene, produced the syndrome analogous to infection by any of several viroids. When one healthy half-leaf was pricked with a needle moistened in 0.2 *M* ethephon, necrotic spots similar to the lesions of a hypersensitive reaction were produced on the treated half-leaf without reaction on the half-leaf treated only with the carrier solution, 0.1% Tween® 80. Analyses of protein extracts from ethylene-induced symptomatic tissues by both SDS and native gels (Figure 10) produced patterns virtually identical to the proteins induced in CEV-infected tomato tissues. However, one protein (C_{10}) appeared to be stimulated only in CEV-infected tissues. The molecular weight of the proteins ranged from 13,500 to 24,600 daltons as determined by migration in SDS gels of protein

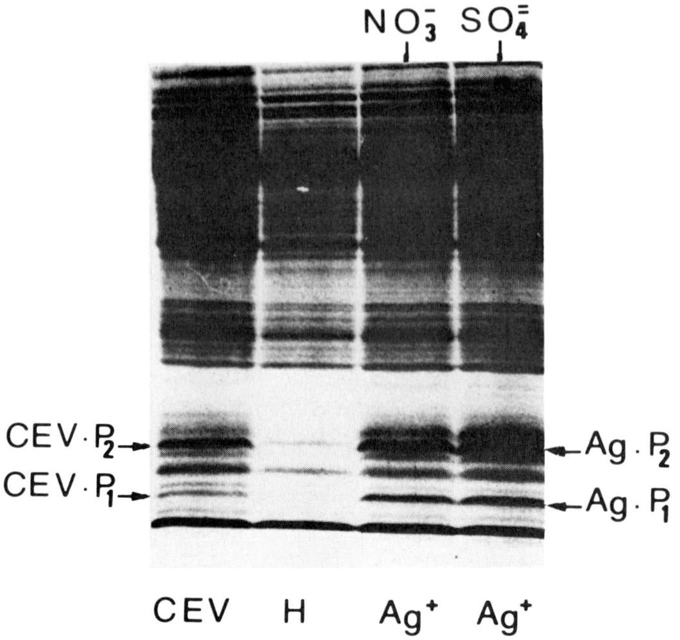

FIGURE 12. SDS-polyacrylamide slab gel (14%) electrophoresis pattern of proteins from *Gynura aurantiaca* DC plants which were either healthy (H), citrus exocortis viroid-infected (CEV), or treated with the indicated compounds. The cation concentration in both Ag^+ treatments was 2.94 m*M* (equivalent to the Ag^+ content in 500 mg/ℓ of $AgNO_3$). Migration of P_1 (13,000 daltons) and P_2 (18,000 daltons) are indicated.

isolated from native gels. The production of protein alterations was localized to treated half-leaves with no evidence for systemic translocation of the stimulus or the proteins.

Metal ions have been demonstrated to produce some metabolic responses in plant tissues that simulate alterations accompanying virus infection. These responses usually result from localized damage or imminent death of affected cells. Foliar spraying of *Gynura aurantiaca* with an aqueous solution of Ag^+ (500 to 1000 mg/ℓ) produced a syndrome of general stunting and leaf epinasty and rugosity (Figure 11) which closely resembled infection by CEV.[94] The effect, which remained localized to treated tissues, resulted in the production of the same PR proteins in as described both *Gynura* and tomato. No viroid-like syndrome was produced by nitrate salts of Cu^{2+}, Zn^{2+}, Cp^{2+}, or Pb^{2+}, whereas both the nitrate and sulfate salts of Ag^+ were equally effective in inducing symptoms (Figure 12). Although it has been reported that Ag^+ at low concentration (100 μg/ℓ) inhibited the production of ethylene by pea tissues[95] and counteracted the effects of ethylene by retarding epicotyl growth, Ag^+ at higher concentrations actually enhanced ethylene production in *Gynura*.

The commonality of host expression resulting from the action of a biological agent, the viroid, as well as in response to even simpler chemical molecules, ethylene and Ag^+, reinforces the generalized nature of the host reactions initiated. At what point convergence of these processes from the divergent inducers or "signals" occurs remains one of the key questions in this topic of viroid pathogenesis. Each additional fragment of correlative data, however, contributes to our best, though still admittedly incomplete, perspective. A major distinction to bear in mind, however, is that the viroid disease response is generated by an intracellular agent that is systemically transported, whereas the ethylene and Ag^+-induced responses are highly localized. In addition, the basic distinction between the localized viroid-induced PR proteins and the systemically induced PR proteins of the more classical model of Van Loon[88] should be recognized. Furthermore, even with their defined association with systemic acquired resistance, the precise biological role of PR proteins themselves has not been determined.

C. Phytohormone Metabolism

From the outset of this discussion of viroid pathogenesis, we have focused on the probable importance of hormonal metabolism on the gross growth properties which can result from viroid infection. Although the growth alteration observed in many viral infections can be related to changes in phytohormone levels, the commonality of symptom expression in most viroid infections suggests the potential of a more direct cause-effect relationship. The observation of common protein profile alterations induced by both viroid infection and the phytohormone, ethylene, now simply strengthens this possibility. Even in advance of this observation, studies were made of the variations in phytohormone content between healthy and viroid-infected tissues, as well as the differential response of these tissues to exogenously applied phytohormones.

Although it can be difficult to correlate data describing detection of endogenous hormone levels in CEV-infected citrus trees in the field[96] and in greenhouse-grown *Gynura aurantiaca* containing CEV,[97] results from both studies indicated a net decrease in gibberellin. Application of exogeneous gibberellic acid$_3$ (GA_3) to infected *G. aurantiaca* resulted in the typical reaction of internode elongation. Therefore, the diminished levels of GA may be related to the stunting reaction. Analyses of auxin-like substances and abscisic acid (ABA) in the two diverse tissue sources produced contradictory results which may be attributed to selection of sampled materials. In infected citrus, auxin-like substances decreased and an inhibitor presumed to be abscisic acid increased, whereas in *Gynura,* IAA-like auxin remained unchanged, except for the appearance of low levels of an additional auxin-like substance, and ABA was not affected by the presence of viroid. It does not seem surprising that detectable alterations occur in endogenous hormone concentrations. However, establishment of the significance of these patterns in viroid pathogenesis requires supporting data from additional CEV hosts and/or other host-viroid combinations as well as evaluation of parameters other than fresh weight.

When Flores and Rodriguez[98] investigated the phenomenon of reduced root initiation in CEV-infected *Gynura,* they also observed a two- to threefold reduction in diffusible auxin-like substances from apical buds, which would appear to be in conflict with the analyses of endogenous auxin previously reported.[97] Although the differential rooting response was dramatic, these studies were initiated with cuttings from *Gynura* plants made 30 days after onset of symptoms when severe growth retardation of the shoots was already evident. Therefore, the reduction of rooting could simply reflect a delay in root initiation as a result of a generally diminished metabolic capacity of viroid-infected plants, and not necessarily specific limitation of production of substances like auxin. Flores and Rodriguez[98] have suggested factors affecting auxin transport or increased oxidase activities in diseased tissues as potentially primary causes of the effect. A similar delay in root initiation has also been described for CSV-infected chrysanthemum,[99] as well as the almost complete inhibition of rooting from CEV-infected tomato.[98] Also, the reduction in root systems and the formation of "rake-like" branching in HSV-infected hop plants[100] support the concept of a generalized growth response.

An alternative approach to explore the question of phytohormone aberrations in viroid-infected tissues was reported by Duran-Vila and Semancik.[101] In order to minimize any effects of differential transport between healthy and diseased tissues, leaf disks, stem segments, and callus cultures were cultivated in vitro under different hormone regimes. In this manner, not only the effect of individual phytohormones, but also combinations thereof, can be tested, an important advantage since the balance between components such as auxin and cytokinins is critical to tissue response. The hybrid tomato (*Lycopersicon esculentum* × *Lycopersicon peruvianum*) used in these studies displays an indeterminant growth pattern and thus can provide a continuous source of uniform tissues of more moderate symptoms than *L. esculentum* cv. Rutgers when inoculated with CEV.

Healthy leaf disks or stem segments of the hybrid tomato showed maximum root differentiation in 1 week when cultured on a medium containing 1 mg/ℓ of IAA, whereas CEV-infected tissues displayed a 1 to 2 week delay in initiation of roots which never exceeded 2 to 3 mm in length.[101] Increased IAA concentration (up to 50 mg/ℓ) or prolonged exposure did not enhance root production from infected tissues, whereas healthy tissues reacted with shorter roots and discoloration. This differential rooting response was also observed in disorganized callus generated from the two tissue types. A similar effect was produced with leaf discs and stem segments exposed to other auxins, including 2,4-dichlorophenoxyacetic acid (2,4-D), naphthalene acetic acid (NAA), and indolebutyric acid (IBA).

The significance of this observation is heightened by the absence of any differential response of healthy and CEV-infected tissues to cytokinins (kinetin, 6-benzylaminopurine) or gibberellin.[38] When a kinetin concentration gradient and IAA at 1 mg/ℓ were provided in the same medium, rooting of healthy tissues became inhibited at 0.1 mg/ℓ kinetin. However, under no conditions was the rooting of CEV-infected tissues enhanced.

These studies suggest the existence of some aberration in auxin utilization by CEV-affected tissues and/or a morphological impairment of the tissues preventing the expected reponses to auxin. The inability to induce any alteration in rooting of infected tissues in response to increasing cytokinin concentration further suggests the lack of excessive levels of auxins in diseased tissues which might account for the root differentiation inhibition. Since translocation problems would not provide a logical explanation in this system, interference with auxin recognition or inactivation of auxin by viroid-infected tissues would seem most probable. It is interesting to note under in vitro conditions that leaf disk expansion of healthy tissues exceeded that of infected tissues by at least twofold, possibly suggesting an IAA-related structural anomaly in the cell wall structure of the viroid-containing tissue.[102] Therefore, some combination of factors affecting cell wall structures as well as hormonal metabolism would seem a reasonable projection for a cause of at least one end point of viroid pathogenesis.

The reported[103] lack of response of tissues to auxin following treatment with ethylene and the concomitant increase in hydroxyproline-rich cell wall proteins (extensin) presents an attractive relationship. Whether or not these factors are, indeed, integrated in a cause and effect relationship or simply may be implicated as constituents of the secondary cascade of biological events resulting from viroid infection will remain a recurrent theme in the following sections as we develop this overview.

The enhanced production of ethylene has been associated with the induction of symptoms resulting from invasion of pathogenic biological agents.[104] This stimulation can then be related to some of the demonstrated effects of ethylene which also describe viroid diseases, such as severe stunting, epinasty, chlorosis, premature senescence, and inhibition of rooting. We have already observed that plants treated with the ethylene-producing agent, ethephon, responded with the production of proteins identical to those induced in viroid-infected tissues. Therefore, it is not altogether surprising that analyses of ethylene evolution from various viroid-affected tissues have indicated a significant stimulation in ethylene evolution. Organized tissues in the form of either cuttings, leaf disks, or stem apices of *Gynura aurantiaca*,[171] or tomato plantlets in culture[38] displayed a ratio of ethylene evolution from CEV-infected to healthy tissue of from 1.2 to 2.1. This value was markedly increased (1.8 to 3.8) when disorganized tissues, callus, or suspension cells in culture were tested. It should be re-emphasized that these cultures represented nonsymptomatic tissues, since callus derived from healthy or CEV-infected tissues could not be differentiated; however, the CEV-containing materials were believed to represent a more homogeneous sampling of infected cells than organized plant tissue. Therefore, a direct correlation can be drawn between increased ethylene evolution and, not only symptom expression, but also the presence of the viroid.

Expansive ramifications are possible in any explanations of symptom expression in diseased plants with which ethylene is implicated.[95,104,105] Gross symptom effects, including retardation of hypocotyl elongation, epinasty, and development of chlorosis have been attributed to the action of ethylene. The concomitant increase in 1-aminocyclopropane-1-carboxylic acid (ACC) as the immediate precursor of ethylene has been linked to the stimulation of ethylene production.[106] It is instructive to consider possible sites of control in this pathway as contained in the model of Mattoo and Anderson.[107] Following the ''cell perception'' of an external ''stimulus'' by the plasma membrane, the *de novo* synthesis of ACC synthase might be initiated by some ''signal''. Whether the viroid-RNA might function either directly as such a ''stimulus'' or ''signal'' in ethylene induction or its effects are mediated through some nuclear event is only conjecture.

An important correlation essential to any further development of the concept is the activation of ACC synthase in concert with the expression of viroid pathogenesis. The key position occupied by ethylene in various interactions in metabolism of phytohormones, PR protein, and enzymes such as peroxidase presents constant challenges to determination of the principal cause-effect relationship of this phytohormone in viroid pathogenesis. Nevertheless, one must be taken by the correlative coherency of the fragmentary data collected thus far and the importance of ethylene in the total process. The enhancement of synthesis of hydroxyproline-rich glycoproteins induced by ethylene and its implication on cell wall growth and pathogenic reaction[108] provide a relevant bridge to introduce aberrations in carbohydrate metabolism.

D. Carbohydrate Metabolism and Cell Wall Composition

We have already introduced a number of observations which reinforce the concept that some aspects of carbohydrate metabolism are affected during the process of viroid pathogenesis. One of the most frequently mentioned observations of viroid cytopathology, as noted in Table 2, is the occurrence of some cell wall distortion.

The idea that the kind of structural anomaly morphologically apparent in cells of viroid-infected tissues is reflective of a fundamental lesion in the actual cell wall composition is supported by data of Marton et al.,[20] which describes a differential release of protoplasts from healthy and viroid-infected tissues. It is significant to note in these studies that although the ratio of protoplasts released from infected vs. healthy tissue varied from about 10 to 50% from greenhouse grown plants, the same and even more dramatic response (release rate from infected/healthy cells = 10%) was observed when callus or suspension cells were employed. Therefore, it appears that the putative viroid-induced cell wall lesion is retained as a basic property of the cell in the absence of organized plant growth. Mention of a similar reduction in protoplast yield (50 to 60%) from CEV-infected plants was also made by Mühlbach et al.[14]

To probe the specific nature of this apparent aberration in cell wall composition, β-1,3 glucanases from *Streptomyces* and *Cytophage,* as well as from a commercial source, Zymolyase-5000 (Miles Laboratories), were provided in supplement to the digestion mixture containing macerozyme and cellulase.[20] With this treatment, an enhancement in protoplast release from infected tissues of three- to fivefold was observed, as opposed to less than onefold from healthy tissues. Although the exogenously added glucanases were not highly purified preparations, the common stimulation in protoplast release produced by enzymes from different sources tends to implicate a β-1,3 glucan anomaly in viroid-containing cells.

The accumulation of materials with callose-type linkages is not uncommon to host-pathogen interactions and the stimulation of glycoprotein production has been associated with localized infections.[109] The structural stability of cell walls can be affected by hydrolysis of β-glucans, which occurs during elongation of *Zea mays* coleoptiles.[110] Thus, the presence of higher than usual amounts of these linkages in walls of viroid-containing cells could contribute to restriction of cell wall breakdown during attempted protoplast release. This further introduces the possible association between an RNA (viroid) molecule with cell regulatory activity and complex carbohydrate structural units of the cell wall which also display properties of regulatory molecules.[111]

Additional support for the existence of a cell wall-related dysfunction can be drawn from the reduction in expansive growth demonstrated by viroid-infected tissues cultured in vitro.[38] Whether this response results from a "hardening" of the basic cell wall structure and/or increased cell-cell adhesion is unknown. How these factors interrelate with the phytohormone effects previously described presents a formidable list of qualified candidates for a most complex series of reactions. At the present time, unfortunately, we find ourselves still in the rather primitive phase of data collection wherein we may only hope to correlate available observations with current theories of expansive cell growth.

Nevertheless, we believe a valuable tool for such analyses has been presented in the system for callus and cell suspension culture of viroid-containing cells which apparently retain the cell wall lesion of intact infected plants.[20] Essential to any detailed analysis of morphology, metabolism, or composition is an adequate quantity of materials from which to obtain uniform sampling. Systems which utilize in vitro cell cultures originally derived from one or a very small number of viroid-infected cells may best serve this criterion.

Relevant to specific aspects of carbohydrate metabolism, some preliminary observations have been made with CEV-infected tomato cells. Evaluation of various carbohydrates as sole in vitro carbon sources indicated that both healthy and viroid-infected cells were sustained on media containing only sucrose or glucose, but not when provided with galactose, arabinose, or xylose.[38] However, when reduced concentrations (10 g/ℓ) of these sugars were added in supplement to the normal 30 g/ℓ of sucrose, CEV-infected cells responded with a two- to threefold enhanced growth in cell mass over healthy cells in the presence of arabinose or xylose (Figure 13). This response became more evident when cells were grown in medium containing sucrose supplemented with a gradient of 1 to 30 g/ℓ of either of the pentoses. In the presence of 1 to 10 g/ℓ of xylose the viroid-infected cells displayed little inhibition, as opposed to a decrease of healthy cells to about 20% of maximum growth. Since arabinose and xylose are utilized by *Zea mays* coleoptile sections when supplied in an incubation medium,[112] some alteration in pentose metabolism and not uptake or transport would appear to be indicated by the stimulated growth of CEV-containing cells. It is also interesting to note that in the process of auxin-induced wall loosening, arabinose, xylose, and glucose are the major neutral sugars contributing carbon to the enlarging cells.[113] Thus, major factors involved in the processes of cell wall loosening as well as synthesis appear possibly to be affected during viroid pathogenesis.

Two major questions which remain center on whether the reduction in the expansive growth of viroid-affected tissues can be extrapolated to explain the stunting of intact plants and if this reaction is accompanied by an alteration in the basic cell wall chemical composition. A component considered to be prominent in the control of cell wall extensibility is extensin.[114-116] Accumulation of this insoluble, hyroxyproline-rich glycoprotein accounts for reduction in cell expansive growth by cross-linking to itself via formation of isodityrosine bridges catalyzed by peroxidase.[111] This data suggests two critical components, extensin and peroxidase, as logical subjects of analysis in viroid-affected tissues. It is of interest here to also point out the additional activity of peroxidase in the oxidative inactivation of the phytohormone, IAA.

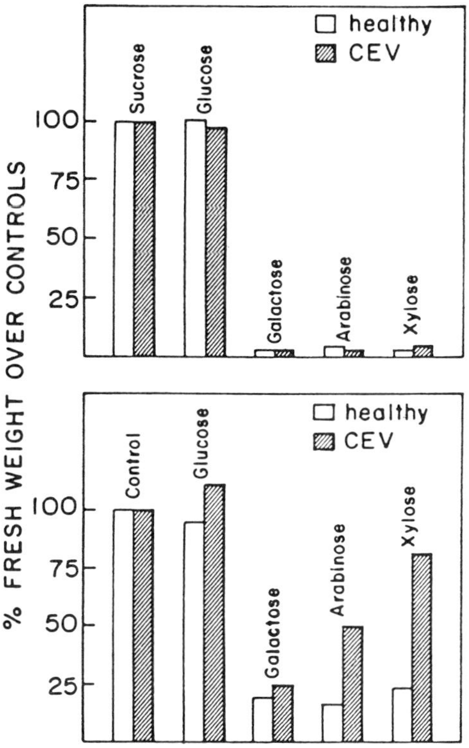

FIGURE 13. Effect of sugars on callus growth. (Upper) Effect of indicated sugars supplied at 30 g/ℓ as the sole carbon source on the growth of healthy and citrus exocortis viroid (CEV)-containing callus. Histograms represent the percentage growth as related to controls cultured in a medium containing sucrose. (Lower) Effect of various sugars indicated, supplemented at 10 g/ℓ to a medium containing 30 g/ℓ of sucrose. Histograms represent the percentage growth as related to controls cultured in medium containing only sucrose.

Employing healthy and CEV-infected tomato suspension cell cultures, Wang et al.[172] have made analyses of purified cell walls. Detection of increased levels of the key constituents of extensin, hydroxyproline, and specific arabinosyl residues[117] can be interpreted as indicating an increase in the extensin content of cell wall preparations. Isolated walls from CEV-infected cells contained a higher proportion of arabinosyl and galactosyl residues. Methylation and amino acid analyses of whole cell wall samples from CEV-infected cells revealed a consistant increase over healthy cell walls in 2,3,5-tri-*o*-methylarabinose (arab-terminal), 3,5-di-*o*-methylarabinose (arab 2), 2,5-di-*o*-methylarabinose (arab 3), and hydroxyproline and serine (Table 5). In addition, dramatic increases were observed in the synthesis of 5-linked arabinosyl (2,3-di-*o*-methyl-arabinose), which can be conjectured to be part of a nonextensin, insoluble arabinogalactan, and in linear β-1,3-linked glucosyl residues (2,4,6-tri-*o*-methyl-glucose). The dramatic increase (two- to tenfold) in β-1,3 glucan deposited on the walls of CEV-containing cells can be correlated with the improved protoplast release from infected cells recorded when β-1,3 glucanase was added to standard cellulase-macerozyme digestion mixtures.[20]

An additional glycoprotein constituent of the cell wall which has not as yet been reported to have been monitored in viroid-affected tissues is peroxidase. Preliminary observations have been made which indicate that the presence of CEV enhances the total peroxidase activity in *Gynura aurantiaca* and tomato and induces the premature appearance of one anionic isoenzyme which possesses both peroxidase and IAA oxidase activities.[173] Also, an

Table 5
RELATIVE MOLAR RATIO OF THE PUTATIVE HYDROXYPROLINE ARABINOSIDES FROM THE WALL GLYCOPROTEIN OF HEALTHY AND CEV-INFECTED, SUSPENSION-CULTURED TOMATO CELLS

	Relative molar ratio (μmol/g dry glycoprotein)[a]	
Compound	Healthy	CEV-infected
Arabinosyl terminus	1.00	1.35
3-Linked arabinosyl	1.00	1.37
2-Linked arabinosyl[b]	1.00	1.30
2-Linked arabinosyl	1.00	1.30
Hydroxyproline	1.00	1.33

[a] The relative abundance from the walls of healthy cells was expressed as 1.00. The dry walls were normalized as pure glycoprotein for this computation.

[b] Two 2-linked arabinosyl residues are the repeating unit of hydroxyproline tetraarabinoside side chains of insoluble extensin.

additional cationic form of a peroxidase isoenzyme has been observed in viroid-infected tomato.[171] Because of the inherently variable nature of this molecule, as evidenced by the wide range of isoenzymes already characterized,[118] as well as its enzyme activity as an IAA oxidase and as a catalyst in formation of interpeptide bridges (isodityrosine) among wall glycoproteins, it seems reasonable that some alteration from the normal peroxidase pattern might be effected in viroid-containing tissues.

When suspension cells like those utilized in analyses of cell wall constituents were inspected by scanning electron microscopy, prominent protrusions of various sizes and forms were apparent on the surface of CEV-infected cells, which contrasted markedly with the essentially smooth healthy cell surface.[41] Although the chemical nature of these sturctures has not yet been determined, the possibility of a relationship of their presence to the cell wall compositional alterations presented here may be entertained. It should also be recalled that the inhibition of expansive growth of leaf discs from CEV-infected plants in culture was accompanied by disorganization of the vascular bundles and a reduction in intercellular spaces, suggesting greater cell-cell adhesion. These observations seem compatible with an altered cell surface quality. Pectic projections have been reported[119] on the surface of wound callus cells of *Daucus carota*. These projections were correlated with cells containing highly convoluted plasmalemmae and multivesicular bodies similar to the plasmalemmasome structures reported in some viroid-infected cells. This tenuous correlation suggests, once again, a homology of cellular reactions in healthy plant cells with the reactions observed in viroid pathogenesis with only, perhaps, an increase in intensity. If a specific restriction in cell wall growth was accompanied by even normal plasma membrane development, one might anticipate observation of an accumulation of plasma membrane as perhaps expressed in the endocytic invaginations (plasmalemmasomes) reported in viroid-affected cells.

IX. INTEGRATION OF METABOLIC AND STRUCTURAL EFFECTS OF VIROID INFECTION

As has already been discussed, PR protein production is related to the developmental syndrome generated as a consequence of the stress-ethylene apparently induced by CEV infection. With this finding, questions arise as to whether (1) any of these proteins play an active or positive role in giving rise to these developmental symptoms or (2) if any of these

PR proteins is a breakdown product produced by degenerative alterations induced in the host proteins.

Relative to the first question is the fact that an ethylene-induced accumulation of hydroxyproline-rich proteins has been found in diseased plants and that this accumulation has been correlated with the developmental alterations produced.[108,120] The idea that hydroxyproline-rich glycoproteins (mainly extensin) together with peroxidases (also glycoproteins) can control the rate of growth and the size and shape of the plant cells and of the whole plant by controlling primary cell wall extension has been recognized.[111] Under this conception, cross-linking in extensin would tighten the cell wall structure, thus restricting cell expansion. The degree of cross-linking in the extensin backbone seems to be controlled through formation of isodityrosyl residues catalyzed by a cell wall-bound peroxidase. According to Eisinger,[121] there is also a soluble cytoplasmic peroxidase which appears to catalyze the hydroxylation of proline-containing peptides and which might influence the synthesis of wall proteins. Thus, the inhibitory effects of ethylene on cell expansion would be explained mainly by the induced enhanced levels of peroxidases, which would increase the content of hydroxyproline-rich proteins in the cell wall, and by the increased degree of cross-linking in extensin.

Could any of the viroid-induced PR proteins fit into the above outlined hypothesis? It is interesting to note that although the glycoprotein- and hydroxyproline-rich natures of some of the cationic PR proteins have not yet been tested, evidence indicates that all viroid PR proteins, except C_9, fulfill the prerequisite of being basic in nature. In addition, two of these proteins, C_4 and C_7, are charged isomers with a molecular weight (36,000 daltons), very close to that (40,000 daltons) reported for the polypeptide component of extensin.[122] Because these proteins are soluble in a buffered aqueous medium (pH 2.8), a relationship may exist with the soluble cytoplasmic protein precursor of extensin. The possibility that some of these cationic PR proteins could be isoenzymes of peroxidase has already been presented.

The alternative possibility that some of these proteins may be degradation products has not yet been tested with the methods used to study native proteins. However, using SDS-PAGE, it has been found[123] that CEV infection induces in *Gynura aurantiaca* plants the appearance of proteolytic activities which are able to produce in healthy extracts the accumulation of two low molecular weight polypeptides which comigrate with the CEV-induced P_1 and P_2. Although the identity of the in vitro proteolytic activity with the native CEV-P_1 and -P_2 polypeptides needs to be proven, these results indicate that P_1 and P_2 might not be native proteins actively synthesized, but rather, proteolytic products. Whether these fragmentary polypeptides play any role in viroid pathogenesis or are merely the end products of degenerative processes accompanying the disease remains to be elucidated.

A formidable cast of compatible "actors" have thus far been assembled. However, it still remains to be determined which is to play the lead and which will be prominent in supporting roles in the production, "Viroid Pathogenesis". Figure 14 presents a schematic attempt to highlight some of the significant metabolic events that have been described in the preceding sections as incited by some "signal" mediated through a nucleus-related function. Rather than attempt to construct a putative pathway or series of direct cause-effect relationships, which would be highly presumptuous at this point with the fragmentary data available, we have presented an array of "cytoplasmic factors" and "wall constituents" which have the ability to interact in the process of growth inhibition and cell wall hardening. The dissimilar size of the two groupings purposely suggests that different interactions may be possible, some with significant impact, others of little consequence. Certainly, even this complexity belies the greater metabolic subtleties that occur during viroid infection. And, as our data base for viroid pathogenesis increases, we may be presented with an even more bewildering collection of "metabolic gears" from which to fashion the correct "timepiece" for viroid disease development.

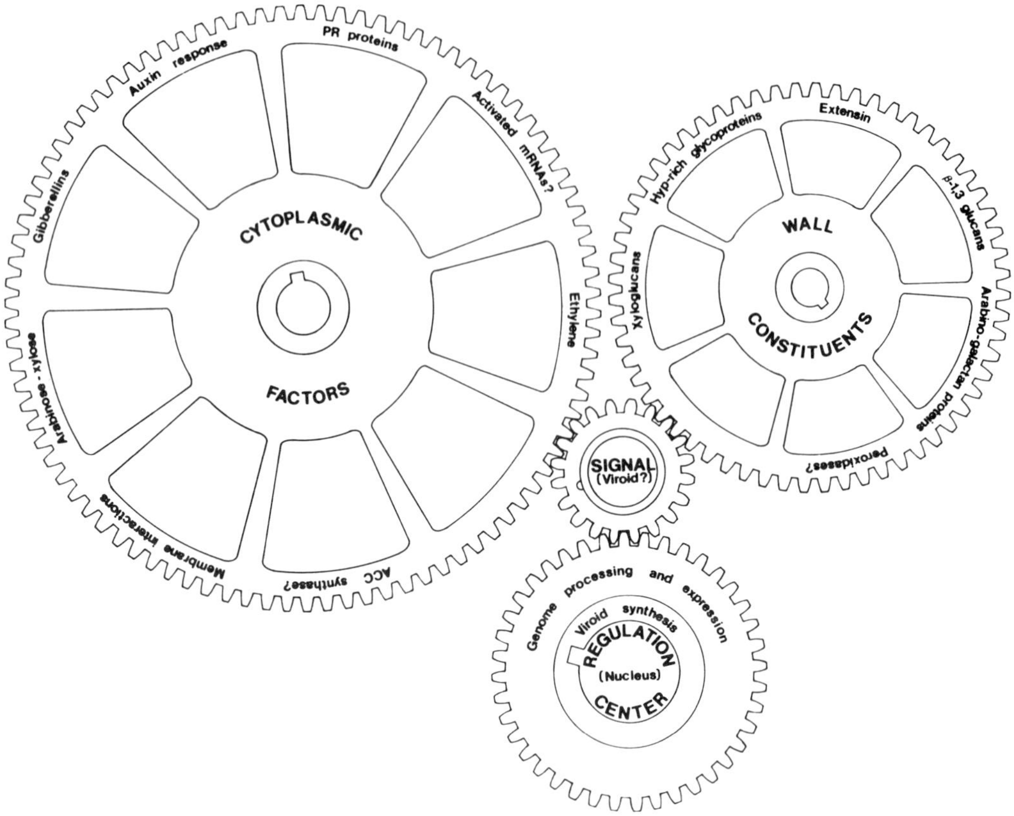

FIGURE 14. Stylized presentation of principal effects and proposed (?) factors involved in viroid pathogenesis.

Although ethylene occupies an equivalent position among the ''cytoplasmic factors'', its role in viroid pathogenesis may very likely be related to the ''transmission'' of viroid effects between the cytoplasm and the cell wall. This concept is supported by the strong correlations in the abilities of both viroids and ethylene to produce certain metabolic effects and similar morphological symptoms. These include the specific stimulation of the production of PR proteins, hydroxyproline-rich cell wall proteins, and probably peroxidases.[124] In addition, both display the capacity to influence β-1,3 glucanase activity,[125,126] and impair auxin transport and giberrellin-stimulated growth.[127-128] A discussion of the complex interactions that occur between ethylene and other plant hormones is beyond the scope of the present topic. Nevertheless, the prime position of ethylene as ''... an important component in the mix of hormonal regulatory factors that control growth, development and senescence''[95] correlates well with our view of its key role in viroid pathogenesis.

Some element of uncertainty is denoted in Figure 14 by (?) as to the importance or detection of viroid-induced alterations, although the inclusion of pre-mRNAs, ACC synthase, and peroxidases seem highly probable at this point. The inclusion of wall constituents such as arabinogalactan proteins (AGPs), which display structural as well as ''informational'' activities, suggests a logical mechanism to explain the modification in cell and tissue development. Experimental verification, however, is essential to support the speculated importance of the AGPs in expressing the identity of tissue and cell types and influencing cell-cell adhesion,[129] and in even more distant extrapolation of the gross effect of stunting of intact plants.

Since complex wall carbohydrates have the potential to act as regulatory molecules involved in controlling gene expression,[130,131] viroid pathogenesis may present the truly unique phenomenon of a regulatory RNA (viroid) originally derived from an exogenous source interacting with or acting upon an endogenous regulatory carbohydrate to effect a persistent morphological alteration or aberration in the host by control of genome expression. In fact, we should at least entertain the possibility that AGPs or other wall constituents which may act as receptors for external ''signals'' between cells may also function in a similar receptor capacity for ''signals'' generated from within cells in response to viroid infection. Alternatively, the viroid itself may have the ability to act as the ''signal''. In either case, with the acceptance that viroid synthesis occurs in the nucleus, it is implicit that these metabolic events are under some nuclear control process, at least at the outset, with the cell subjected to what might be called an internal stress or inducer in the form of the viroid. The precise nature of the ''signal'' or ''keystone interaction'' which sets this intricate and interfacing series of metabolic events on an altered developmental course is of fundamental importance. What has become apparent in this discussion is the host-specified nature of the metabolic parameters affected by viroid infection. What remains most elusive is the nature of the primary reaction between the host processes and the foreign controlling element, the viroid-RNA.

From the persistent nature of the effect of viroid pathogenesis, it would seem logical that this principal event occurs either (a) during a phase in ''genome processing and expression'' at the nuclear level or (b) at a key step in a fundamental metabolic pathway at some extranuclear locus. Even though Figure 14 suggests a predominant flow in the course of events, the distinct opportunity for the action of feedback processes or induction of nucleus-controlled processes by cell wall and membrane components, informational carbohydrates, or cytoplasmic pre-mRNAs through other ''signal'' apparatuses must be entertained. The hypothesis presented by Zimmern[132] for the origin of viroids and RNA viruses and the exchange of genetic information in cells is compatible with this concept. Also, in many ways, our current knowledge allows us to view viroid pathogenesis as analogous to or mimicking a type of systemic host response which normally results in resistance to pathogens. This view becomes even more attractive when we consider the compatibility of the mechanism proposed by Zimmern[132] as well as Ryan[133] for the induction of proteinase inhibitors in plant leaves as a natural protection system. What can be proposed from these diverse phenomena is that viroid pathogenesis could proceed through introduction of (1) a ''signal'' which is recognized by some (2) receptor or target within the host, resulting in (3) the appearance of a host-specified product which is usually under nuclear control.

These perspectives should stimulate and encourage the pursuit of experimentally generated data to challenge not only these views, but correlations made to date simply on the basis of sequence homology. At the outset, it should be interesting to simply challenge the well-characterized in vitro reactions involved in all phases of host genome processing and expression by inclusion of viroid-RNA. Any indication of biological activity, either in stimulation or inhibition by the viroid-RNA in such systems, could at least function to direct our thinking. This approach must still be complemented by an era of continued data collection on such topics as the nature of the viroid in vivo and its implicit intermolecular associations, as well as characterization and determination of compositional alterations of host components, such as the cell walls, in response to viroid.

Many of the challenges issued at the time the existence of the unique small pathogenic RNAs in plants was recognized more than a decade ago still remain. We are fortunate to have available to us potent information on the basic structure of viroids. However, as in most research activities of a biological nature, the physical-chemical characterization of specific molecules is less demanding than the deciphering of complex interactions of biologically active molecules. We are, nevertheless, encouraged to believe that comprehension of the cause-effect relationship resulting in viroid pathogenesis can be achieved because of the relative simplicity inherent in the interaction between a sequence of barely 350 nucleotides and an organism. What is even more exciting is the potential prize, that in deciphering the intricacies of host-viroid interactions, knowledge of a unique controlling process, perhaps functional in all biological systems, may be revealed.

A major quandary in viroid research centers on whether these pathogenic RNAs are restricted to the plant kingdom. If the primary metabolic impact of viroid infection is focused on cell walls in the production of the diseased condition that has attracted our attention, it remains for us to first determine the biological action of similar molecules in animal systems before their existence as animal pathogens can be verified. What form this expression might take in the absence of cell walls remains a fundamental consideration.

X. ANTIVIROID AND PROTECTION PHENOMENA

A. Developmental Resistance

Throughout this discussion of viroid pathogenesis we have encountered data at random indicating the restriction of viroid infection. Such restriction usually has resulted from some limitation in the initiation of primary infection rather than the development of viroid-free tissues from diseased plants, although the success of shoot tip grafting confirms the efficacy of the latter procedure. Here, we wish to direct our attention to instances of host-viroid interactions which suggest the presence of conditions for the rejection or accommodation of the infectious viroid-RNA. Even in the application of shoot tip culturing techniques, we might question whether the apparent freedom from viroid of the terminal 0 to 0.5 mm portion of the apex results from the absence of a vascular connection and the concomitant slow penetration of the viroid or from some cellular incompatibility for the viroid-RNA. It would seem only equitable that since we have already proposed the existence of ''target'' cells characterized by some mechanism of developmental susceptibility, we should at least entertain the possibility of viroid resistance imparted by some condition or process that occurs prior to such a stage in these same cells. In other words, a developmentally governed ''window'' of susceptibility may exist.

In the infection of *Gynura aurantiaca* with CEV we have observed a condition which truly might be regarded as an expression of ''developmental resistance''. Not only is the primary infection restricted to the terminal apex as in other expressions of viroid infection, but the first leaf region displaying symptoms, as well as viroid accumulation and synthesizing capacity, is many times accompanied by a connected half-leaf of apparently totally unaffected cells.[11] In this half-leaf, as well as in all more mature leaves, the viroid is not able to initiate infection, but does successfully invade and affect axial buds that emerge from the base of these leaves.

Another characteristic of these symptomatic half-leaves and the initially infected apex is the stimulation of host-specified PR proteins[10] analogous to the PR proteins associated with acquired systemic resistance in the hypersensitive reaction of tobacco inoculated with TMV.[104] Although there is some evidence against the direct role of tobacco PR proteins in the development of the so-called acquired systemic resistance,[134] their possible involvement cannot be definitively excluded. The question then arises as to whether the CEV-induced PR proteins are associated with resistance reactions of this type. Protecting phenomena against CEV infection in *Gynura aurantiaca* plants have been stimulated by spraying the plants with aqueous solutions of 100 ppm $AgNO_3$ (Figure 15) or 480 ppm of ethephon.[135] Analogous results were obtained after stimulating a hypersensitive reaction by pricking leaves with a needle moistened with 0.2 *M* ethephon in a similar way as the induction of systemic resistance in tobacco Samsun NN against TMV was performed by Van Loon.[104,136]

Unlike the situation reported in the case of acquired systemic resistance in tobacco plants, no alteration of the normal SDS-PAGE protein pattern nor the production of any PR protein that could be related to the induced systemic CEV resistance was detected when young upper

FIGURE 15. Protective effect of Ag^+ applied at low concentration (100 mg/ℓ) against infection of *Gynura aurantiaca* DC by citrus exocortis viroid (CEV). Treated plants were sprayed every 5 days with an aqueous solution of $AgNO_3$ with 0.1% Tween® 80. Healthy (left) and CEV-infected (middle) plants are displayed with Ag^+-treated (right) plant.

leaves of protected *Gynura* plants were analyzed. The fact that tomato PR proteins are neither systemically induced nor translocated[171] is consistent with this finding. However, the possibiltiy that some of the PR proteins may be involved in defense mechanisms against other pathogens should still be entertained. In this respect, it is pertinent to note that as a consequence of the stress imposed by CEV infection and, potentially, by a number of different agents, ethylene mediates a cascade of events which produce a developmental syndrome. At the same time, the potential defenses of the plant are enhanced due to the effects of ethylene. However, but defenses associated with ethylene production are of different types. They can be of a specific nature such as the ethylene-mediated production of phytoalexins and the ethylene-induced chitinase against bacterial and fungal plant pathogens.[120,130,137] There are also less specific mechanisms of defense, as seems to be the case of the protecting reactions stimulated by ethylene against viroid infection. Such reactions can be considered examples of acquired systemic resistance. Although CEV- and ethylene-stimulated PR proteins seem not to be directly related with systemic acquired resistance, some of these proteins might be formed as part of a general response to stress-induced ethylene, but function significantly in alternative, specific defense mechanisms.

Related to this possibility, we might speculate that some of the viroid-induced PR proteins might be similar to ethylene-stimulated chitinase[130] or β-1,3 glucanase activities.[126] The high isoelectric points and reported molecular weights of some of the PR proteins induced in tomato by CEV are compatible with this idea.[137] Interestingly, one of these PR proteins (C_8) has a molecular weight (30,000 daltons) coincident with that reported for chitinase. The fact that the production of both chitinase and β-1,3-glucanase are mediated by ethylene strengthens this possibility.

The similarities in the types of reactions and their products associated with protection mechanisms in plants and with normal senescence are significant to the discussion of viroid pathogenesis because they reinforce the critical nature of the stage of host plant development in the initiation of viroid infection. Although the protection reactions are induced by one or more exogenous agents, both phenomena are under host-specified control and the senescence process occurs naturally in the absence of initiating pathogens.

Temporal considerations, therefore, weigh heavily in the success or failure of host-viroid interactions. Relevant to this time dependence is the reported occurrence of ''cross-protection'' among viroids, defined as ''... the interference in symptom expression'' by Niblett et al.[138] In this study, initial inoculation of a mild isolate of PSTV ''cross-protected'' against a challenge inoculation by a severe isolate of PSTV, as well as against CEV in tomato, while in chrysanthemum both CSV and PSTV protected against CEV. The effect produced consisted essentially of a delay of symptom expression similar to the protection reaction noted with the previously described pretreatment with ethephon or Ag^+. It seems probable that a number of biological as well as chemical agents may induce the same or similar host-mediated response, which essentially mimics a state of premature senescence. This perspective helps to explain the absence of a strict homology among viroids in what has been reported as ''cross-protection'' in viroid infections. The practical implication of this topic will be further developed in Chapter 4, which deals with the agricultural applications of viroid ''infection''.

Returning to the theme of developmental resistance, a further example to consider is the strikingly severe zonal chlorosis which results on leaves of avocado displaying the ''shock phase'' symptoms of ASV (Figure 16). We have already commented that the bulk of viroid accumulation[43] as well as synthetic activity[168] occurs in the heavily chlorotic or bleached areas. Therefore, the viroid appears to be excluded from the green areas of heavily symptomatic leaves and a rather distinct interface zone of demarcation from the bleached regions becomes apparent, even in the absence of a hypersensitive or necrotic reaction. For this reason, we have already introduced the possibility that the initiation of viroid infection is somehow governed by developmental events in the cells of the marginal meristem, which is, in effect, another aspect of developmental resistance. In the absence of confirming experimental data we should also entertain the alternative possibility that the discretely delimited zone in the ASV infected leaves could result from the presence of an antiviroid chemical barrier which restricts the successful spread of viroid-RNA.

A somewhat analogous condition could be artificially induced in PSTV-infected tomato, wherein albinism was observed in the new growth from plants held under continuous light at 30°.[49] A three- to tenfold increase in viroid concentration occurred in the white tissue. This effect, presumably specific to viroid-infected plants, could only be produced on tissues not fully developed prior to exposure to the specific light/temperature regime, indicating a correlation of the ability of viroid to induce a ''symptom'' expression with a developmental stage of the host. Furthermore, the absence of vein necrosis in white leaves of diseased plants grown under continuous light suggested a condition that might favor the spread of the viroid which under normal conditions of light and temperature would be restricted by such necrosis.

B. The Symptomless Carrier Condition

The occurrence of seed transmission among viroids is unpredictable, with both high and low levels of transmission reported for PSTV and no evidence of seed transmission reported for CEV. In avocado infected with ASV, a curious dual type of seed transmission has been reported by Wallace and Drake.[29] A low frequency of seed transmission was recorded from symptom-expressing trees. Virtually 100% seed transmission was possible from infected but symptomless trees or branches from symptomatic trees. The seedlings and their succeeding

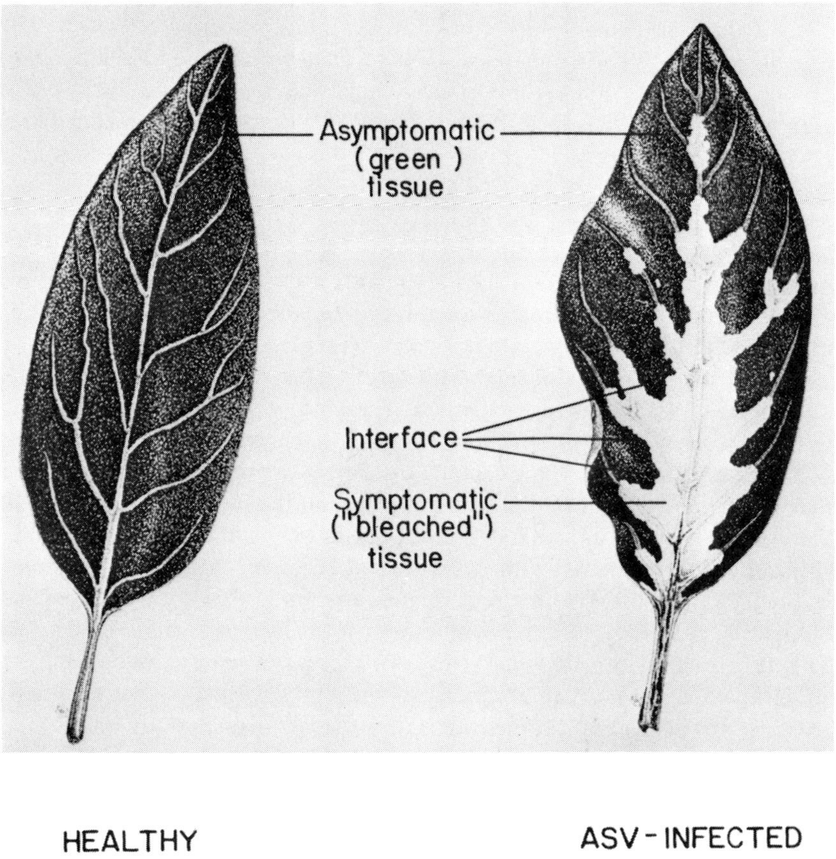

FIGURE 16. Healthy (left) and avocado sunblotch viroid-infected (right) avocado leaves diagramed. Sectoring of nonexpressing and severely "bleached" or chlorotic areas as delimited by the non-necrotic interface.

generations "carried" but did not express symptoms of ASV, although buds excised from them could transmit the symptomatic condition when grafted into healthy plants. The description of such infected plants as "symptomless carrier" (SC) plants was adopted. This phenomenon suggests that host cells usually susceptible to viroid infection can under some circumstances "learn" to live with the viroid or adapt the viroid to become recognized as a nondisruptive component of the host nucleic acid population. The persistence of CEV in tomato cell culture without expression of detrimental symptoms also indicates a high degree of coordination of viroid with host nucleic acid synthesis and regulation.[41]

From the different durations of incubation periods in transmission tests it was suggested that a lower concentration of ASV occurred in the symptomless carrier plants.[29] More recent analyses[168] indicate, however, that in fact a higher concentration of viroid occurs in the SC tissue than in the green (asymptomatic) areas of symptom expressing tissues. Therefore, a distinction among viroid distributions can be made on the basis of symptom expression, which also provides an explanation for the "protection" of SC tissues from challenge inoculation. If the viroid present in the symptomless carrier condition becomes associated with host tissues in a uniform and perhaps totally systemic manner, germination of seed produced under these conditions could initiate new cell division in the presence of the viroid-RNA without the antagonistic encounter that apparently occurs when extraneous viroid-RNA is introduced into a fully developed plant. Therefore, no shock reaction or symptom expression results. The existence of latent or symptomless conditions in the presence of viroid-like RNA has led to acceptance of the concept that viroid synthesis need not be inexorably linked to pathogenesis. What is not presently known is whether an alteration in the basic structure of the viroid occurs during this process. In any case, a most intriguing phenomenon is witnessed when an infectious foreign RNA can be introduced and perpetuated in apparent harmony with host metabolism.

XI. BENEFICIAL ASPECTS OF VIROID INFECTION

Given these demonstrations of the feasibility of host and viroid compatibility, it is possible to develop the theoretical consideration of the beneficial aspects of viroid-altered growth and development. In the viroid, we are presented with (1) a nuclear agent that (2) not only persists, but is regulated in equilibrium with host nucleic acids as a small nuclear RNA, and (3) the effects of which can be moderated. Therefore, the viroid would seem a suitable candidate for development as an effective, exogenously introduced regulatory molecule or an extra-chromasomal nuclear factor capable of altering growth responses; i.e., we should consider the potential use of the viroids in parallel with the more fashionable prospects of introducing foreign genes in an integrated fashion. Not only is the transmission and survival of viroid-like RNAs enhanced by their nuclear association, but the expression of their biological potential may prove to be less complex or cryptic than the expression of foreign genes. Minimally, it would seem reasonable to at least attempt to harness the potential as expressed by viroid-RNA for the control of some aspects of plant development. The demonstrated ability to limit the natural spread of the viroid, which is not vectored by any biological agent other than man, further encourages this pursuit.

The possible use of viroid in inducing stunting in tree crops to enhance yield and facilitate harvesting is already in the testing stage and will be addressed in Chapter 4. More speculative ventures can be projected from some of the observations made on viroid-containing cells in culture, including their survival under (1) high temperature stress and (2) high hormone (herbicide) levels, as well as their potential for (3) prolonged viability. These factors, coupled with the differential sensitivity between healthy and viroid-infected cells to pentoses, already suggest the feasibility for development of a selective medium for growth of viroid-containing cells in support of eventual development of plants bearing altered traits imparted by the viroid. Finally, we have presented a relationship wherein the presence of viroid apparently can induce enhancement of host cell wall hardening, which could potentially be expolited to reduce the effectiveness of fungal and bacterial plant pathogens that must overcome this physical barrier. Whether the experimental observations can be functionally extrapolated to whole plant systems of agricultural importance presents a formidable but worthwhile challenge.

XII. CORRELATIONS AND SPECULATION ON THE MOLECULAR INDUCTION OF VIROID PATHOGENESIS

Included in each of the substantial number of reviews describing viroid research is usually a section addressing the question of viroid pathogenesis. For the most part, these comprise very brief statements as to the position of viroids as abnormal regulatory RNA functioning through some mechanism involving an imbalance of growth hormone activity to produce a variety of symptoms resembling those of typical virus infections. We have, I believe, attempted to make a departure from this course in our efforts (1) to focus on commonality of the biological activities expressed among various viroids and (2) to offer the perspective that the viroid may mimic a normal class of regulating RNA that induces abnormal metabolism insofar as it is out of phase with the normal processes of plant growth and development.

Some attempts have been made to relate what is known of viroid structures and replication to the process of pathogenesis. Caution is necessary in these correlations, however, since adequate evidence has been presented for the absence of linkage between viroid synthesis and viroid pathogenesis. For example, the suggestion that viroid pathogenesis may evolve from the generalized sequestering of RNA polymerase II (Pol II)[139] should predict the production of a dramatic effect on the synthesis of total mRNA in viroid infected plants. No experimental data in support of this contention has yet been offered, other than the in vitro indication that binding of viroid RNA to Pol-II-rich preparations is possible. Similar studies testing the interaction between viroid-RNA and Pol-I and -III might prove to sustain this suggestion or challenge the specificity of the binding.

Certainly, the greatest activity to date related to possible molecular bases of viroid pathogenesis centers on the correlation of base sequence homologies between viroid-RNA and known small nuclear RNAs. Once again, these exercises must be viewed with some circumspection, since the well characterized snRNAs are derived from mammalian systems and yet to be described from plant systems. Furthermore, the function of most of the snRNAs remains unclear.

A key role in the processing of eukaryotic mRNA has been suggested for the U1a-RNA identified as a 165 nucleotide snRNA.[140] From a homology between one pentanucleotide and one hexanucleotide sequence in PSTV with the 5′ end of rat U1-RNA, a model was proposed by Dickson[141] for the possible involvement of viroids in the mRNA splicing process. It was suggested that viroid-RNA could contain base sequences complementary to the intervening sequence (intron) region of plant pre-mRNAs, thus providing a possible mechanism for inducing a pathogenic expression by replacing or competing for the U1-RNA. Extrapolating upon this suggestion, Diener[142] proposed a " . . . presumed intron ancestor of the viroid component". An additional note of caution in this correlation seems warranted, since it has been noted that ASV can be viewed either as not containing similar homologous sequences in the viroid or the complementary sequence[143] or containing such regions of homology,[144] depending upon the detection parameters employed.

An even more generalized phylogenetic relationship has been suggested between viroid-RNA and snRNAs of mammalian origin, based on computer-generated base sequence homology analyses.[144,145] It was observed that more than 60% of the sequence of U5A RNA was homologous with ASV and 81% of the entire U3B RNA from Novikoff hepatoma cells was contained in PSTV. Although the authors contended that the left half of the circular PSTV molecule may be phylogenetically related especially to the U3 snRNA, they were cautious in noting the possible existence of structural convergence as a requirement for biosynthesis. The fact that the positions of the two major areas of homology were included within the central conserved region of PSTV would tend to justify this qualifying statement. An additional region of homology with U3B RNA located in the left side of the PSTV sequence (nucleotides 34 to 52) includes a sequence (nucleotides 43 to 56) that has been suggested as a possible center at which molecular conformation and pathogenicity are related.[2,146] The fact that this region is identical in a severe isolate of CEV which displays dwarfing and presumably rootstock scaling as well as in one which induces only the dwarfing reaction[147] suggests that this putative relationship may not be adequate to explain all expressions of pathogenesis, especially in different host-viroid combinations.

Yet another instance of structural homology has been suggested between viroids in the PSTV group and the ends of transposable genetic elements and retroviral proviruses.[148] The base sequence array of the five viroids in this group, PSTV, CEV, TASV, and CSV share the similar properties of (1) a polypurine region in the left-hand region of the molecule, (2) an inverted imperfectly repeated sequence in the central conserved region which (3) is flanked by the dinucleotides U-G...C-A. In addition to suggesting that these characteristics may indicate a similar function or common ancestry, the authors acknowledge that the similarities may be "fortuitous". These varied hypotheses on the origin and/or function of viroids drawn from sequence homologies share one common feature, the yearning for experimental confirmation.

XIII. SUMMATION

In conclusion, we first must plea for absolution from our sins of perhaps a too impassioned and, at times, biased perspective. Ours has not been an intent to convince, so much as to stimulate and challenge our own best instincts and limited abilities.

Upon the solid framework of the primary structure of the viroids, as representatives of a specialized class of small nuclear RNAs, we have attempted to decorate with a ''functional cloth''. What emerges at present, however, is at best a patchwork not yet entirely compatible with the substance and potential of the biological activity of these most unique agents. We are confronted with the most intriguing mechanisms for the viroid as a regulator or effector of a ''class'' of common responses. Future interpretations may see the convergence onto a view of the viroid as a foreign regulatory RNA acting upon constitutive regulating components of its host in a most direct cause-effect relationship. The actions of viroids upon host growth and development processes suggest an unusual coordination and dependence. This view offers the viroid as an investigative probe to normal host processes and, at the same time, questions whether ''disease'' is a natural or inevitable consequence of the presence of viroid-RNA. From this we might project that given a better comprehension of the biological potential of viroids, a novel mechanism for altering plant growth expression may evolve: a wonderful enterprise and an encouraging perspective.

ACKNOWLEDGMENTS

We acknowledge the collective communicative effort stimulated by the emergence of viroids, not only to the associates, past and present, of our laboratories, but all those contributors brought together here who have been moved to direct a portion of their valued existence to this focus.

And in the end, we must acknowledge that ours has hardly been ''discovery,'' any more than the searching Spaniards discovered the magnificence of Tenochtitlan, but rather that we have been touched by the exhilaration of exploring that which was already before us and yet unseen, that which is our environment but is not our realm. For, however we may feel in momentary conquest or control, that which reveals still destines us to a subjugative role. We remain captured by the powers and intricacies of a most profound and intriguing natural order which sustains us, while at the same time challenges our vision. And our reward is in knowledge and insight, which themselves must reflect a form of sublime revelation.

REFERENCES

1. **Semancik, J. S.,** Small pathogenic RNA in plants — the viroids, *Ann. Rev. Phyopathol.*, 17, 461, 1979.
2. **Sänger, H. L.,** Biology, structure, function and possible origin of viroids, in *Encyclopedia of Plant Physiology,* Vol. 14B, Parthier, B. and Boulter, D., Eds., Springer-Verlag, Basel, 1982, 368.
3. **Diener, T. O.,** *Viroids and Viroid Diseases,* Wiley Interscience, New York, 1979.
4. **Diener, T. O.,** Viroids, *Trends Biol. Sci.*, 4, 133, 1984.
5. **Riesner, D., Colpan, M., Goodman, T. C., Nagel, L., Schumacher, J., and Steger, G.,** Dynamics and interactions of viroids, *J. Biomol. Struct. Dyn.*, 1, 669, 1983.
6. **Schlemmer, A., Roistacher, C. N., and Semancik, J. S.,** A unique infectious RNA associated with citron showing symptoms typical of citrus exocortis disease, *Phytopathology,* 75, 946, 1985.
7. **Roistacher, C. N., Gumpf, D. J., Nauer, E. M., and Gonzales, R.,** Cachexia disease: virus or viroid, *Citrograph,* 68, 111, 1983.
8. **Koganezawa, H., Yanase, H., and Sakuma, T.,** Viroid-like RNA associated with apple scar skin (or dapple apple) disease, *Acta Hort.*, 130, 193, 1983.

9. **Paliwal, Y. C. and Singh, R. P.**, Cytopathological changes induced by potato spindle tuber viroid in *Scopolia sinensis, Can. J. Bot.*, 59, 677, 1981.
10. **Conejero, V. and Semancik, J. S.**, Exocortis viroid: alteration in the proteins of *Gynura aurantiaca* accompanying viroid infection, *Virology*, 77, 221, 1977.
11. **Semancik, J. S. and Vanderwoude, W. J.**, Exocortis viroid: cytopathic effects at the plasma membrane in association with pathogenic RNA, *Virology*, 69, 719, 1976.
12. **Lyndon, R. F.**, The shoot apex, in *Cell Division in Higher Plants*, Yeoman, M. M., Ed., Academic Press, New York, 1976, 285.
13. **Semancik, J. S., Grill, L. K., and Civerolo, E. L.**, Accumulation of viroid RNA in tumor cells after double infection by *Agrobacterium tumefaciens* and citrus exocortis viroid, *Phytopathology*, 68, 1728, 1978.
14. **Mühlbach, H. P., Camacho-Henriquez, A., and Sänger, H. L.**, Isolation and properties of protoplasts from leaves of healthy and viroid-infected tomato plants, *Plant Sci. Lett.*, 8, 183, 1977.
15. **Mühlbach, H. P. and Sänger, H. L.**, Multiplication of cucumber pale fruit viroid in inoculated tomato leaf protoplasts, *J. Gen. Virol.*, 35, 377, 1977.
16. **Stange, L.**, Cellular interactions during early differentiation, in *Encyclopedia of Plant Physiology*, Vol. 17, Linskens, H. F. and Heslop-Harrison, J., Eds., 1984, Springer-Verlag, Basel, 1984, 424.
17. **Yeoman, M. M.**, The mitotic cycle in higher plants, in *The Cell Cycle*, John, P. C. L., Ed., Cambridge University Press, New York, 1981, 161.
18. **Schroeder, C. A.**, Growth and development of the Fuerte avocado fruit, *Am. Soc. Hort. Sci.*, 61, 103, 1953.
19. **Kojima, M., Murai, M., and Shikata, E.**, Cytopathic changes in viroid-infected leaf tissues, *J. Fac. Agric. Hokkaido Univ.*, 61, 219, 1983.
20. **Marton, L., Duran-Vila, N., Lin, J. J., and Semancik, J. S.**, Properties of cell cultures containing the citrus exocortis viroid, *Virology*, 122, 229, 1982.
21. **Murashige, T., Bitters, W. P., Rangan, T. S., Nauer, E. M., Roistacher, C. N., and Holliday, P. B.**, A technique of shoot apex grafting and its utilization towards recovering virus-free citrus clones, *Hort. Sci.*, 7, 118, 1972.
22. **Navarro, L., Roistacher, C. N., and Murashige, T.**, Improvement of shoot-tip grafting *in vitro* for virus-free citrus, *J. Am. Soc. Hort. Sci.*, 100, 471, 1975.
23. **Stace-Smith, R. and Mellor, F. C.**, Eradication of potato spindle tuber virus by thermotherapy and axillary bud culture, *Phytopathology*, 60, 1857, 1970.
24. **Lizarraga, R. E., Salazar, L. F., Roca, W. M., and Schilde-Rentachler, L.**, Elimination of potato spindle tuber viroid by low temperature and meristem culture, *Phytopathology*, 70, 754, 1980.
25. **Kryczynski, S.**, Transmission of viroids and viruses by tissue implantation and transport across the callus barrier, *Phytopathol. Z.*, 106, 63, 1983.
26. **Schumacher, J., Sänger, H. L., and Riesner, D.**, Subcellular localization of viroids in highly purified nuclei from tomato leaf tissue, *EMBO J.*, 2, 1549, 1983.
27. **Fernow, K. H., Peterson, L. C., and Plaisted, R. J.**, Spindle tuber virus in seeds and pollen of infected potato plants, *Am. Potato J.*, 47, 75, 1970.
28. **Singh, R. P. and Finnie, R. E.**, Seed transmission of potato spindle tuber metavirus through the ovule of *Scopolia sinensis, Can. Plant Dis. Surv.*, 53, 153, 1973.
29. **Wallace, J. M. and Drake, R. J.**, A high rate of seed transmission of avocado sun-blotch virus from symptomless trees and the origin of such trees, *Phytopathology*, 52, 237, 1962.
30. **Fawcett, H. S. and Klotz, L. J.**, Bark shelling of trifoliate orange, *Calif. Citrograph*, 33, 230, 1948.
31. **Schneider, H.**, Pathological anatomies of citrus affected by virus diseases and apparently-inherited disorders and their use in diagnoses, *Proc. 1st Int. Citrus Symp.*, 3, 1489, 1969.
32. **Childs, J. F. L., Norman, G. G., and Eichhorn, J. L.**, A color test for exocortis infection in *Poncirus trifoliata, Phytopathology*, 48, 426, 1958.
33. **Fudl-Allah, A., Calavan, E. C., and Desjardins, P. R.**, Comparative anatomy of healthy and exocortis virus-infected citron plants, *Phytopathology*, 61, 990, 1971.
34. **Calavan, E. C., Frolich, E. F., Carpenter, J. B., Roistacher, C. N., and Christiansen, D. W.**, Rapid indexing for exocortis of citrus, *Phytopathology*, 54, 1359, 1964.
35. **Weathers, L. G., Greer, F. C., Jr., and Harjung, M. K.**, Transmission of exocortis virus of citrus to herbaceous plants, *Plant Dis. Rep.*, 51, 868, 1967.
36. **Mohamed, N. A. and Thomas, W.**, Viroid-like properties of an RNA species associated with sunblotch disease of avocados, *J. Gen. Virol.*, 46, 157, 1980.
37. **Momma, T. and Takahashi, T.**, Ultrastructure of hop stunt viroid-infected leaf tissue, *Phytopathol. Z.*, 104, 211, 1982.
38. **Duran-Vila, N.**, Studies of Viroid Pathogenesis: Responses of CEV-Containing Tissues and Cells, Ph.D. dissertation, University of California, Riverside, 1982.
39. **Lawson, R. H. and Hearon, S. S.**, Ultrastructure of chrysanthemum stunt virus-infected and stunt-free mistletoe chrysanthemum, *Phytopathology*, 61, 653, 1971.

40. **Momma, T. and Takahashi, T.,** Cytopathology of shoot apical meristem of hop plants infected with hop stunt viroid, *Phytopathol. Z.*, 106, 272, 1983.
41. **Lin, J. J.,** Coordination Between Host Nucleic Acid Metabolism and Citrus Exocortis Viroid Turnover, Ph.D. dissertation, University of California, Riverside, 1984.
42. **Da Graca, J. V. and Martin, M. M.,** Ultrastructural changes in avocado leaf tissue infected with avocado sunblotch, *Phytopathol. Z.* 102, 185, 1981.
43. **Semancik, J. S. and Desjardins, P. R.,** Multiple small RNA species and the viroid hypothesis for the sunblotch disease of avocado, *Virology*, 104, 117, 1980.
44. **Wahn, K., Rosenberg-de Gomez, F., and Sänger, H. L.,** Cytopathic changes in leaf tissue of *Gynura aurantiaca* infected with the viroid of citrus exocortis disease, *J. Gen. Virol.*, 49, 355, 1980.
45. **Tu, J. C. and Hiruki, C.,** Electron microscopy of cell wall thickening in local lesions of potato viruses in infected red kidney bean, *Phytopathology*, 61, 862, 1971.
46. **Hari, V.,** Ultrastructure of potato spindle tuber viroid-infected tomato leaf tissue, *Phytopathology*, 70, 385, 1980.
47. **Kim, K. S. and Fulton, J. P.,** Plant virus-induced cell wall overgrowth and associated elaboration, *J. Ultrastruct. Res.*, 45, 328, 1973.
48. **Lafleche, D., Bové, C., Dupont, G., Mauches, C., Astier, T., Garnier, M., and Bové, J. M.,** Site of viral-RNA replication in the cells of higher plants: TYMV-RNA synthesis on the chloroplast outer membrane system, in *RNA Viruses/Ribosomes*, North-Holland, Amsterdam, 1972, 43.
49. **Yang, T. C. and Hooker, W. J.,** Albinism of potato spindle tuber viroid-infected Rutgers tomato in continuous light, *Am. Potato J.*, 54, 519, 1977.
50. **Galindo, J., Smith, D. R., and Diener, T. O.,** Etiology of *Planta macho*, a viroid disease of tomato, *Phytopathology*, 73, 49, 1982.
51. **Diener, T. O.,** Potato spindle tuber virus: a plant virus with properties of a free nucleic acid. III. Subcellular location of PSTV-RNA and the question of whether virions exist in extracts or *in situ*, *Virology*, 43, 75, 1971.
52. **Sänger, H. L.,** An infectious and replicating RNA of low molecular weight: the agent of the exocortis disease of citrus, *Adv. Biosci.*, 8, 103, 1972.
53. **Takahashi, T., Yaguchi, S., Oikawa, S., and Kamita, N.,** Subcellular location of hop stunt viroid, *Phytopathol. Z.*, 103, 285, 1982.
54. **Semancik, J. S., Tsuruda, D., Zaner, L., Geelen, J. L. M. C., and Weathers, L. G.,** Exocortis disease: subcellular distribution of pathogenic (viroid) RNA, *Virology*, 69, 669, 1976.
55. **Semancik, J. S., Flores, R., Harper, K., and Lin, J. J.,** Some properties of citrus exocortis viroid (CEV) synthesis in cell-free extracts and cell suspension cultures, in *Plant Infection Agents*, Robertson, H. D., Howell, S. H., Zaitlin, M., and Malmberg, R. L., Eds., Cold Spring Harbor Laboratory, Cold Spring Harbor, N.Y., 1983, 149.
56. **Flores, R. and Semancik, J. S.,** Properties of a cell-free system for synthesis of citrus exocortis viroid, *Proc. Natl. Acad. Sci. U.S.A.*, 79, 6285, 1982.
57. **Semancik, J. S. and Harper, K. L.,** Optimal conditions for cell-free synthesis of citrus exocortis viroid and the question of specificity of RNA polymerase activity, *Proc. Natl. Acad. Sci. U.S.A.*, 81, 4429, 1984.
58. **Zaitlin, M. and Beachy, R. N.,** The use of protoplasts and separated cells in plant virus research, *Adv. Virus Res.*, 19, 1, 1974.
59. **Raychaudhuri, S. P.,** Plant viruses in tissue culture, *Adv. Virus Res.*, 12, 175, 1966.
60. **Ingram, D. S.,** Growth of biotrophic parasites in tissue culture, in *Physiological Plant Pathology*, Heitefuss, R. and Williams, P. H., Eds., Springer-Verlag, Berlin, 1976, 743.
61. **Zelcer, A., Van Adelsberg, J., Leonard, D. A., and Zaitlin, M.,** Plant cell suspension cultures sustain long-term replication of potato spindle tuber viroid, *Virology*, 109, 314, 1981.
62. **Mühlbach, H. P. and Sänger, H. L.,** Continuous replication of potato spindle viroid (PSTV) in permanent cell culture of potato and tomato, *Biosci. Rep.*, 1, 79, 1981.
63. **Mühlbach, H. P., Faustmann, O., and Sänger, H. L.,** Conditions for optimal growth of a PSTV-infected potato cell suspension and detection of viroid-complementary longer-than-unit-length RNA in these cells, *Pl. Mol. Biol.*, 2, 239, 1983.
64. **Hildebrandt, A. C.,** Single cell culture, protoplasts and plant viruses, in *Applied Fundamental Aspects of Plant Cell, Tissues and Organ Culture*, Reinert, J. and Bajaj, Y. P. S., Eds., Springer-Verlag, Berlin, 1977, 581.
65. **Muir, W. H., Hildebrandt, A. C., and Riker, A. J.,** Plant tissue cultures produced from single isolated cells, *Science*, 119, 877, 1954.
66. **Hanson, A. J. and Hildebrandt, A. C.,** The distribution of tobacco mosaic virus in plant callus cultures, *Virology*, 28, 15, 1966.
67. **Sänger, H. L. and Ramm, K.,** Radioactive labelling of viroid-RNA, in *Modification of the Information Content of Plant Cells*, Markham, R., Davies, D. R., Hapwood, D. A., and Horne, R. W., Eds., North-Holland, Amsterdam, 1975, 229.

68. **Schlesinger, M. J., Aliperti, G., and Kelley, P. M.,** The response of cells to heat shock, *Trends Biol. Sci.*, 4, 221, 1982.
69. **Kanabus, J., Pikaard, C. S., and Cherry, J. H.,** Heat shock proteins in tobacco cell suspension during growth cycle, *Plant Physiol.*, 75, 639, 1984.
70. **Hadidi, A. and Diener, T. O.,** *De novo* synthesis of potato spindle tuber viroid as measured by incorporation of ^{32}P, *Virology*, 78, 99, 1977.
71. **Bastin, M. and Kaesberg, P.,** Radioactive labelling of brome mosaic virus, *J. Gen. Virol.*, 26, 321, 1975.
72. **Morris, T. J.,** Evidence for a single infectious species of potato spindle tuber viroid, *Intervirology*, 11, 89, 1979.
73. **Davies, J. W., Kaesberg, P., and Diener, T. O.,** Potato spindle tuber viroid. XII. An investigation of viroid RNA as a messenger for protein synthesis, *Virology*, 61, 281, 1974.
74. **Hall, T. C., Wepprich, R. K., Davies, J. W., Weathers, L. G., and Semancik, J. S.,** Functional distinctions between the ribonucleic acids from citrus exocortis viroid and plant viruses: cell-free translation and aminoacylation reactions, *Virology*, 61, 486, 1974.
75. **Semancik, J. S., Conejero, V., and Gerhart, J.,** Citrus exocortis viroid: survey of protein synthesis in *Xenopus laevis* oocytes following addition of viroid RNA, *Virology*, 80, 218, 1977.
76. **Matthews, R. E. F.,** Are viroids negative-strand viruses?, *Nature (London)*, 276, 850, 1978.
77. **Grill, L. K. and Semancik, J. S.,** RNA sequences complementary to citrus exocortis viroid in nucleic acid preparations from infected *Gynura aurantiaca*, *Proc. Natl. Acad. Sci. U.S.A.*, 75, 896, 1978.
78. **Haseloff, J. and Symons, R. H.,** Chrysanthemum stunt viroid: primary sequence and secondary structure, *Nucleic Acids Res.*, 9, 2741, 1981.
79. **Zaitlin, M. and Hariharasubramanian, V.,** A gel electrophoretic analysis of protein from plants infected with tobacco mosaic and potato spindle tuber viruses, *Virology*, 47, 296, 1972.
80. **Conejero, V., Picazo, I., and Segado, P.,** Citrus exocortis viroid (CEV): protein alterations in different hosts following viroid infection, *Virology*, 97, 454, 1979.
81. **Flores, R., Chroboczek, J., and Semancik, J. S.,** Some properties of the CEV-P_1 protein from citrus exocortis viroid-infected *Gynura aurantiaca* DC, *Physiol. Plant Pathol.*, 13, 193, 1978.
82. **Camacho-Henriquez, A. and Sänger, H. L.,** Gel electrophoretic analysis of phenol-extractable leaf proteins from different viroid/host combinations, *Arch. Virol.*, 74, 167, 1982.
83. **Camacho-Henriquez, A., Lucas, J., and Sänger, H. L.,** Purification and biochemical properties of the pathogenesis-related protein p^{14} from tomato leaves, *Neth. J. Plant Pathol.*, 89, 308, 1983.
84. **Camacho-Henriquez, A. and Sañger, H. L.,** Analysis of acid-extractable tomato leaf proteins after infection with a viroid, two viruses and a fungus and partial purification of the "pathogenesis-related" protein p^{14}, *Arch. Virol.*, 74, 181, 1982.
85. **Antoniw, J. F., Ritter, C. E., Pierpont, W. S., and van Loon, L. C.,** Comparison of three pathogenesis-related proteins of two cultivars of tobacco infected with TMV, *J. Gen. Virol.*, 47, 79, 1980.
86. **Van Loon, L. C.,** Specific soluble leaf proteins in virus-infected tobacco plants are not normal constituents, *J. Gen. Virol.*, 30, 375, 1976.
87. **Van Loon, L. C., Ed.,** Proceedings of the Workshop on Pathogenesis-Related (b) Proteins in Plants, *Neth. J. Plant Pathol.*, 89, 239, 1983.
88. **Van Loon, L. C.,** Pathogenesis-related proteins, *Plant Mol. Biol.*, 4, 111, 1985.
89. **Reisfeld, R. A., Lewis, V. J., and Williams, D. J.,** Disc electrophoresis of basic proteins and peptides on polyacrylamide gels, *Nature (London)*, 195, 156, 1962.
90. **Davis, B. J.,** Disc electrophoresis. II. Method and application to human serum proteins, *Ann. N.Y. Acad. Sci.*, 121, 404, 1964.
91. **Ferguson, K. A.,** Starch-gel electrophoresis: application to the classification of pituitary proteins and polypeptides, *Metabolism*, 3, 985, 1964.
92. **Carr, J. P.,** Translational control of pathogenesis-related protein synthesis, *Neth. J. Plant Pathol.*, 89, 311, 1983.
93. **Carr, J. P., Antoniw, J. F., White, R. F., and Wilson, T. M. A.,** Latent messenger RNA in tobacco *(Nicotiana tabacum)*, *Biochem. Soc. Trans.*, 10, 353, 1982.
94. **Conejero, V.,** Stimulation of a viroid-like syndrome and inhibition of viroid symptom expression by Ag^+ ion, 5th Int. Congr. Virol., Strasbourg, France, August 2-7, P 21/03, p. 228, 1981.
95. **Lieberman, M.,** Biosynthesis and action of ethylene, *Ann. Rev. Plant Physiol.*, 30, 533, 1979.
96. **Hanks, R. W. and Feldman, A. W.,** Changes in amounts of auxin-like growth promoter, gibberellin, and inhibition in citrus infected with exocortis virus, in Proceedings of the 5th Conference of the International Organization of Citrus Virologists, Price, W. C. Ed., University of Florida Press, Gainesville, 1972, 244.
97. **Rodriguez, J. L., Garcia-Martinez, J. L., and Flores, R.,** The relationship between plant growth substance content and infection of *Gynura aurantiaca* D.C. by citrus exocortis viroid, *Physiol. Plant Pathol.*, 13, 355, 1978.
98. **Flores, R. and Rodriguez, J. L.,** Altered pattern of root formation on cuttings of *Gynura aurantiaca* infected by citrus exocortis viroid, *Phytopathology*, 71, 964, 1981.

99. **Horst, R. K., Langhans, R. W., and Smith, S. H.,** Effects of chrysanthemum stunt, chlorotic mottle, aspermy and mosaic flowering and rooting of chrysanthemums, *Phytopathology,* 67, 9, 1977.
100. **Yaguchi, S. and Takahashi, T.,** Survival of hop stunt viroid in the hop garden, *Phytopathol. Z.,* 109, 32, 1984.
101. **Duran-Vila, N. and Semancik, J. S.,** Effects of exogenous auxins on tomato tissue infected with the citrus exocortis viroid, *Phytopathology,* 72, 777, 1982.
102. **Cleland, R.,** Auxin and H^+ excretion: the state of our knowledge, in *Plant Growth Substances,* Proc. 10th Intl. Conf. Plant Growth Substances, Madison, Wisc., Skoog, F., Ed., Springer-Verlag, Berlin, 1979, 71.
103. **Sadava, D. and Chrispeels, M. J.,** Hydroxyproline-rich cell wall protein (extensin): role in the cessation of elongation in excised pea epicotyls, Dev. Biol., 30, 49, 1973.
104. **Van Loon, L. C.,** Regulation of pathogenesis by symptom expression in diseased plants by ethylene, in *Ethylene: Biochemical, Physiological and Applied Aspects,* Fuchs, Y. and Chalutz, E., Eds., Martinus Nijhoff/Dr. W. Junks Publishers, The Hague, 1984, 171.
105. **Bell, A. A.,** Biochemical mechanism of disease resistance, *Ann. Rev. Plant Physiol.,* 32, 21, 1981.
106. **Adams, D. O. and Yang, S. F.,** Ethylene biosynthesis: identification of 1-aminocyclopropane-1-carboxylic acid as an intermediate in the conversion of methionine to ethylene, *Proc. Natl. Acad. Sci. U.S.A.,* 76, 170, 1979.
107. **Mattoo, A. K. and Anderson, J. D.,** Wound-induced increase in 1-aminocyclopropane-1-carboxylate synthase activity: regulatory aspects and membrane association of the enzyme, in *Ethylene: Biochemical, Physiological and Applied Aspects,* Fuchs, Y. and Chalutz, E., Eds., Martinus Nijhoff/Dr. W. Junks Publishers, The Hague, 1984, 139.
108. **Esquerré-Tugayé, M. T., Lafitte, C., Mazau, D., Toppan, A., and Touzé, A.,** Cell surfaces in plant-microorganism interactions. II. Evidence for the accumulation of hydroxyproline-rich glycoproteins in the cell wall of diseased plants as a defense mechanism, *Plant Physiol.,* 64, 320, 1979.
109. **Brown, R. G. and Kimmins, W. C.,** Hypersensitive resistance. Isolation and characterization of glycoproteins from plants with localized infection, *Can. J. Bot.,* 51, 1917, 1973.
110. **Huber, D. J. and Nevin, D. J.,** β-D-Glucan hydrolase activity in *Zea* coleoptile cell walls, *Plant Physiol.,* 65, 768, 1980.
111. **McNeil, M. M., Darvill, A. G., Fry, S. C., and Albersheim, P.,** Structure and function of the primary cell walls of plants, *Ann. Rev. Biochem.,* 53, 625, 1984.
112. **Carpita, N. C., Brown, R. A., and Weller, K. M.,** Uptake and metabolic fate of glucose, arabinose, and xylose by *Zea mays* coleoptiles in relation to cell wall synthesis, *Plant Pathol.,* 69, 1173, 1982.
113. **Darvill, A. G., Smith, C. J., and Hall, M. A.,** Cell wall structures and elongation growth in *Zea mays* coleoptile tissue, *New Phytol.,* 80, 503, 1978.
114. **Lamport, D. T. A.,** The protein component of primary cell walls, *Adv. Bot. Res.,* 2, 151, 1965.
115. **Cleland, R.,** Distribution and metabolism of protein-bound hydroxyproline in an elongating tissue, the *Avena* coleoptile, *Plant Pathol.,* 43, 865, 1968.
116. **Cleland, R. and Karlness, A. M.,** A possible role of hydroxyproline-containing proteins in the cessation of cell elongation, *Plant Physiol.,* 42, 669, 1967.
117. **Akiyama, Y. and Kato, K.,** Hydroxyproline arabinosides from suspension-cultured tobacco cell wall, *Agric. Biol. Chem.,* 40, 2343, 1976.
118. **Farkas, G. L. and Stahmann, M. A.,** On the nature of changes in peroxidase isoemzymes in bean leaves infected by southern bean mosaic virus, *Phytopathology,* 56, 669, 1966.
119. **Davies, W. P. and Lewis, B. G.,** Development of pectic projections on the surface of wound callus cells of *Daucus carota* L., *Ann. Bot.,* 47, 409, 1981.
120. **Boller, T.,** Ethylene-induced biochemical defenses against pathogens, in *Plant Growth Substances,* Wareing, P. F., Ed., Academic Press, New York, 1982, 303.
121. **Eisinger, W.,** Regulation of pea internode expansion by ethylene, *Ann. Rev. Plant Physiol.,* 34, 225, 1983.
122. **Stuart, D. A. and Varner, J. E.,** Purification and characterization of a salt extractable hydroxyproline-rich glycoprotein from aerated carrot discs, *Plant Physiol.,* 66, 787, 1980.
123. **Conejero, V., Segado, P., Bellés, J. M., and Granell, A.,** Citrus exocortix viroid (CEV): new data regarding the low molecular weight polypeptides associated with viroid infection, *Neth. J. Plant Pathol.,* 89, 308, 1983.
124. **Ridge, I. and Osborne, D. J.,** Regulation of peroxidase activity by ethylene in *Pisum sativum:* regulation of ethylene, *J. Exp. Biol.,* 21, 843, 1970.
125. **Wong, Y. S. and Maclachlan, G. A.,** 1,3-β-D-Glucanases from *Pisum sativum* seedlings, *Plant Physiol.,* 65, 222, 1980.
126. **Abeles, F. B. and Forrence, L. E.,** Temporal and hormonal control of β-1,3 glucanase in *Phaseolus vulgaris* L., *Plant Physiol.,* 45, 396, 1970.
127. **Palmer, J. H. and Halsall, D. M.,** Effects of transverse gravity stimulation, gibberellin and indoleacetic acid upon polar transport of IAA C^{14} in the stem of *Helianthus annus, Physiol. Plant.,* 22, 59, 1969.

128. **Stewart, R. N., Lieberman, M., and Kunishi, A. T.,** Effects of ethylene and gibberellic acid on cellular growth and development in apical and subapical regions of etiolated pea seedlings, *Plant Physiol.,* 54, 1, 1974.
129. **Fincher, G. B., Stone, B. A., and Clarke, A. E.,** Arabinogalacton-proteins: structure, biosynthesis, and function, *Ann. Rev. Plant Physiol.,* 34, 47, 1983.
130. **Boller, T., Gehri, A., Mauch, F., and Voegeli, V.,** Ethylene induced chitinase, is it a pathogenesis-related protein?, *Neth. J. Plant Pathol.,* 89, 318, 1983.
131. **Burg, S. P.,** Ethylene in plant growth, *Proc. Natl. Acad. Sci. U.S.A.,* 70, 591, 1973.
132. **Zimmern, D.,** Do viroids and RNA viruses derive from a system that exchanges genetic information between eukaryotic cells?, *Trends Biol. Sci.,* 7, 205, 1982.
133. **Ryan, C. A.,** Proteinase inhibitors in plant leaves: a biochemical model for pest-induced natural plant protection, *Trends Biol. Sci.,* 3, 148, 1978.
134. **Fraser, R. S. S. and Clay, C. M.,** Pathogenesis-related proteins and acquired systemic resistance: causal relationship or separate effects?, *Neth. J. Plant Pathol.,* 89, 283, 1983.
135. **Conejero, V.,** Do viroids elicit host-pre-existing mechanisms of response? Abstr. 4th Int. Conf. on Comparative Virology, Banff, Alberta, Canada, nr. W 12-1, p. 192, 1982.
136. **Van Loon, L. C.,** Induction by 2-chloroethylphosphonic acid of viral-like lesions, associated proteins and systemic resistance in tobacco, *Virology,* 80, 417, 1977.
137. **Abeles, F. B., Basshart, R. P., Forrence, L. E., and Habig, W. H.,** Preparation and purification of glucanase and chitinase from bean leaves, *Plant Physiol.,* 47, 129, 1970.
138. **Niblett, C. L., Dickson, E., Fernow, K. H., Horst, R. K., and Zaitlin, M.,** Cross protection among four viroids, *Virology,* 91, 198, 1978.
139. **Rackwitz, H. R., Rohde, W., and Sänger, H. L.,** DNA dependent RNA polymerase II of plant origin transcribes viroid RNA into full-length copies, *Nature (London),* 291, 297, 1981.
140. **Lerner, M. R., Boyle, J. A., Mount, S. M., Wolin, S. L., and Steitz, J. A.,** Are snRNPs involved in splicing?, *Nature (London),* 283, 220, 1980.
141. **Dickson, E.,** A model for the involvement of viroids in RNA splicing, *Virology,* 115, 216, 1981.
142. **Diener, T. O.,** Are viroids escaped introns?, *Proc. Natl. Acad. Sci. U.S.A.,* 78, 5014, 1981.
143. **Symons, R. H.,** Avocado sunblotch viroid-primary sequence and proposed secondary structure, *Nucleic Acids Res.,* 9, 285, 1981.
144. **Kiss, T. and Solymosy, F.,** Sequence homologies between a viroid and a small nuclear RNA (snRNA) species of mammalian origin, *FEBS Lett.,* 144, 318, 1982.
145. **Kiss, T., Posfai, J., and Solymosy, F.,** Sequence homology between potato spindle tuber viroid and U3B snRNA, *FEBS Lett.,* 163, 217, 1983.
146. **Flores, R.,** Is the conformation of viroids involved in their pathogenicity?, *J. Theor. Biol.,* 108, 519, 1984.
147. **Visvader, J. E. and Symons, R. H.,** Comparative sequence and structure of different isolates of citrus exocortis viroid, *Virology,* 130, 232, 1983.
148. **Kiefer, M. C., Owens, R. A., and Diener, T. O.,** Structural similarities between viroids and transposable elements, *Proc. Natl. Acad. Sci. U.S.A.,* 80, 6234, 1983.
149. **Chen, W., Tien, P., Zhu, Y. X., and Lin, Y.,** Viroid-like RNAs associated with burdock stunt disease, *J. Gen. Virol.,* 64, 409, 1983.
150. **Hollings, M. and Stone, O. M.,** Some properties of chrysanthemum stunt, a virus with the characteristics of an uncoated ribonucleic acid, *Ann. Appl. Biol.,* 74, 333, 1973.
151. **Diener, T. O. and Lawson, R. H.,** Chrysanthemum stunt: a viroid disease, *Virology,* 51, 94, 1973.
152. **Semancik, J. S. and Weathers, L. G.,** Exocortis virus of citrus: associated infectivity with nucleic acid preparations, *Virology,* 36, 326, 1968.
153. **Semancik, J. S. and Weathers, L. G.,** Exocortis disease: evidence for a new species of "infectious" low molecular weight RNA in plants, *Nature (London),* 237, 242, 1972.
154. **Van Dorst, H. J. M. and Peters, D.,** Some biological observations on pale fruit, a viroid-incited disease of cucumber, *Neth. J. Plant Pathol.,* 80, 85, 1974.
155. **Sasaki, M. and Shikata, E.,** On some properties of hop stunt disease agent, a viroid, *Proc. Jpn. Acad., Ser B,* 53, 109, 1977.
156. **Diener, T. O. and Raymer, W. B.,** Potato spindle tuber virus: a plant virus with properties of a free nucleic acid, *Science,* 158, 378, 1967.
157. **Diener, T. O.,** Potato spindle tuber virus. IV. A replicating, low molecular weight RNA, *Virology,* 45, 411, 1971.
158. **Walter, B.,** Un viroide de la tomate en Afrique d l'Ouest: identité avec de viroide du "potato spindle tuber"?, *CR Acad. Sci. Paris,* 292, 537, 1981.
159. **McLean, A. P. D.,** Bunchy top disease of tomato, *S. Afr. Dept. Sci. Bull.,* 100, 36, 1931.
160. **Thomas, W. and Mohamed, N. A.,** Avocado sunblotch — a viroid disease?, *Aust. Plant Path. Soc. Newslett.,* 1, 2, 1979.

161. **Palukaitis, P., Hatta, T., Alexander, D. McE., and Symons, R. H.,** Characterization of a viroid associated with avocado sunblotch disease, *Virology*, 99, 145, 1979.
162. **Romaine, C. P. and Horst, R. K.,** Suggested viroid etiology for chrysanthemum chlorotic mottle disease, *Virology*, 64, 86, 1975.
163. **Randles, J. W.,** Association of two ribonucleic acid species with cadang-cadang disease of coconut palm, *Phytopathology*, 65, 163, 1975.
164. **Owens, R., Smith, D. R., and Diener, T. O.,** Measurement of viroid sequence homology by hyridization with complementary DNA prepared *in vitro*, *Virology*, 89, 388, 1978.
165. **Shikata, E., Sano, T., and Uyeda, I.,** An infectious low molecular weight RNA was detected in grapevines by molecular hybridization with hop stunt viroid cDNA, *Proc. Jpn. Acad.*, 60(6), 202, 1984.
166. **Flores, R., Duran-Vila, N., Pallas, V., and Semancik, J. S.,** Detection of viroid and viroid-like RNAs from grapevine, *J. Gen. Virol.*, 66, 2095, 1985.
167. **Desjardins, P. and Sasaki, P.,** Personal communication.
168. **Semancik, J. S.,** Unpublished data.
169. **Flores, R.,** Unpublished data.
170. **Zaitlin, M.,** Personal communication.
171. **Granell, A., Bellés, J. M., and Conejero, V.,** Unpublished data.
172. **Wang, M. C., Lin, J. J., Duran-Vila, N., and Semancik, J. S.,** Alteration in cell wall composition and structure in viroid-containing cells, *Physiol. Mol. Pl. Pathol.*, 28, 107, 1986.
173. **Rodriguez, J. L., and Flores, R.,** Unpublished data.

Chapter 4

BIOLOGICAL INTERACTIONS AND AGRICULTURAL IMPLICATIONS OF VIROIDS

S. M. Garnsey and J. W. Randles

TABLE OF CONTENTS

I. VIROIDS AS PLANT PATHOGENS

A. Agricultural Impact

Viroids have become valuable subjects for molecular biological research because of their unique properties. The intense interest in viroids as experimental subjects is quite understandable, but their significance as plant pathogens which have economic impact on agricultural production should not be overlooked. It was actually the disease effects of viroids which stimulated research to determine the cause of these diseases and to develop controls for them.[1,2] These studies ultimately led to the characterization of viroids as a unique new type of pathogen.

Cadang-cadang disease of coconut, potato spindle tuber of potato, and chrysanthemum stunt of chrysanthemum are three examples of viroid diseases which have had major impacts on production of specific crops.[2,3-6] Cadang-cadang disease certainly illustrates the devastating potential of viroids (Figure 1). It has killed about 30 million coconut palms in the past 40 years, and losses of about 500,000 trees per year continue.[6-8] These losses have all occurred within a relatively small area in the Philippines and Guam, and damage will be far greater if coconut industries in other areas are exposed. The chrysanthemum industry of the U.S. was nearly destroyed by chrysanthemum stunt disease within a few years after its first appearance.[4,5,9,10] Potato spindle tuber disease has been a less dramatic, but persistent problem to potato production in the U.S. for over 60 years[2] and has gradually appeared in other areas.[11,12] Avocados, citrus, glasshouse cucumbers, hops, and tomatoes have also suffered viroid disease problems of varying severity in different countries.[2,11,13-16]

FIGURE 1. Coconut palm plantation in the central Philippines severely affected by cadang-cadang disease. Numerous palms are already dead and those remaining are in various states of disease. Symptoms usually do not appear until trees are at least 10 years old. Trees gradually decline and usually die within 8 to 16 years after the first symptoms appear. Tree loss has been estimated at up to 30 million palms, all in a relatively limited area.

Diener[17] has suggested that viroids have become significant plant pathogens largely due to modern agricultural practices, and there is justification for this view. Viroid diseases have appeared in several crops which are grown intensively in essentially monoculture systems. A number of these crop plants are multiplied enormously by vegetative means from limited parental stocks and are exposed to frequent mechanical contact in horticultural operations. This combination of factors is ideal for the creation of viroid disease epidemics.

The origin of viroids remains obscure, but it is clear that they can exist in a variety of plant hosts in essentially a symptomless state.[2,17-19] So far, these latent sources have been exposed mostly by chance when a disease-susceptible host becomes exposed and then infection is spread and subsequently recognized as a disease problem. In at least one case, a latent viroid infection was detected by chance using assay procedures for low molecular weight RNAs.[20] At this point, the extent and distribution of viroids in symptomless wild and cultivated plants remains unknown, but it seems reasonable to assume that only a portion of the total reservoir of viroids which occur in naturally infected plant hosts has been exposed. It is also probable that under appropriate conditions additional viroids will be expressed in agricultural crops and could cause significant injury. This threat can be countered to some extent by greater efforts to screen plants for latent viroid infections using modern assay and molecular probe procedures.[17]

In several instances, well-known viroid diseases are being controlled or suppressed by certification and sanitation programs.[2,5,9,11,19] It can be argued that potato spindle tuber, citrus exocortis, and chrysanthemum stunt diseases are not major agricultural problems now, because they are controlled and crop losses are not extensive. They do remain a constant problem, however, in the sense that the pathogens have not been eliminated; the host plants are still susceptible and production costs are increased to constantly maintain the required control and certification programs. The recent outbreak of potato spindle tuber disease in potato breeding lines in Australia[12] is a reminder that these programs are not infallible.

It is also troublesome that for most viroid diseases the primary source of inoculum remains unknown.[2,6,14,21] Thus, the causal viroid may be controlled in the economic crop affected such as citrus, potatoes, and hops, but it is often uncertain that these crops are the primary reservoir for the viroids which attack them.[17] Spread from other hosts into a target crop is difficult to counteract until the alternate sources are identified. The picture is even less bright for cadang-cadang disease, the most destructive viroid disease described so far. Despite intensive research for many years, the reservoir of the causal viroid and the means of natural spread of the viroid within the target crop remain unknown.[6,7] Control of the recurrent epidemics of cadang-cadang disease by development of tolerant or resistant cultivars will take many years.

The commonly practiced controls for viroid diseases such as certification of propagation stocks and sanitation procedures assume the existence of technical resources and a level of knowledge which may not be affordable or available in the agricultural communities of many undeveloped or impoverished countries. Pressure to rapidly expand production of clonally propagated crops such as potatoes or citrus by introducing new germplasm sources can occur and can lead to disaster unless appropriate safeguards are employed. Unfortunately, the need for safeguards such as certification is not usually recognized until it is too late. When new crops are introduced, the use of disease-free mother stock and a simple certification program can often provide effective and economical disease control, but once an inoculum reservoir is established, the cost and problems of disease control increase.

Viroids are not an overnight curiosity which have been discovered and will simply disappear once interest in their special properties fades. They are a factor which will continue to affect production of certain types of agricultural crops. New viroid-induced problems will continue to appear and will require additional action to prevent crop loss.

B. General Effects

Viroids tend to have somewhat similar effects in different symptomatic hosts.[2,3,11,22-25] Stunting is probably the most common symptom,[11] but one that can readily be overlooked unless appropriate controls are present. Other common effects are necrosis, often as a primary effect in or near leaf veins, various patterns of chlorosis, and distortion of certain plant parts.[2,3,11] Some of these effects are illustrated in Figures 2 to 4. The specific effect on yield varies with the portion of the plant harvested as the crop. In the case of potatoes, the quality (shape) and quantity of tubers is affected by potato spindle tuber disease and causes loss.[2,26] The foliage effects in potato are not directly significant, but if severe presumably would also contribute to yield loss. In the case of chrysanthemum stunt and chrysanthemum chlorotic mottle diseases, it is the effects on foliage and bloom which are significant.[2,4,5,10] Hop stunt disease reduces cone size and quality.[15] Citrus tree size is reduced by exocortis disease (Figure 3), but the quality of the fruit is not affected and yield is only reduced in proportion to reduction in tree canopy (see also Section V).[19,27-30] In the case of cadang-cadang disease, nut shape and size is affected initially (Figure 5), and a permanent loss of production occurs when the palms eventually die.[6,8]

The overall economic severity of a disease is a function of both the severity of effect and the distribution of the disease. Tomato bunchy top disease, for example, causes very severe symptoms on those plants which become infected. but because its incidence has not become frequent and widespread, overall losses have been small.[2] In contrast, potato spindle tuber disease may cause only mild symptoms (depending on the strain and the host),[26,31] but even

FIGURE 2. Chlorotic leaf spotting symptoms of chrysanthemum stunt viroid (CSV) in a sensitive *Chrysanthemum* cultivar. Affected plants and blooms are also usually severely stunted and unsuitable for marketing. (Photo courtesy R. Lawson.)

FIGURE 3. (A) Severely stunted, exocortis-infected sweet orange *(Citrus sinensis)* tree on trifoliate orange *(Poncirus trifoliata)* rootstock. (B) Tree of comparable age without exocortis. Mild isolates cause less severe stunting. Infected trees on exocortis-tolerant rootstocks are not stunted.

FIGURE 4. Fruit symptoms of avocado sunblotch. Similarities to sun scald injury gave rise to descriptive name of this disease. Chlorotic lesions are also produced on twigs of sensitive cultivars. (Photo courtesy H. Ohr.)

small yield reductions per plant are significant if multiplied over the vast acreages which can be afflicted from use of diseased propagation stocks.

Viroids may also infect many plant hosts without causing noticeable symptoms. Latent hosts may be significant in the epidemiology of various viroid diseases and in formulating controls.[17,26]

C. Viroid Diseases

General effects of some viroid diseases have already been indicated in previous sections. Specific symptoms have also been described in some detail in previous publications.[2,3,8,10,13-15,19,30] Our purpose is not to catalog all the possible effects various viroids can induce, but to highlight some of the major disease problems. Table 1 summarizes some of the information and also lists the abbreviations used hereafter for each disease viroid in this chapter.

1. Cadang-Cadang Disease

The coconut cadang-cadang viroid (CCCV) has been the most destructive viroid pathogen to date. As noted earlier, upwards of 30 million trees have been destroyed and losses still continue at the rate of several hundred thousand trees per year.[6-8] The disease has been known for over 50 years, and several epidemics have occurred. Cadang-cadang disease has spread slowly in the central part of the Philippines. The Tinangaja disease of coconuts in Guam is apparently similar,[32] but CCCV has not yet appeared in other coconut growing areas.

The cadang-cadang disease progresses gradually from the onset of first symptoms which include changes in the shape and size of the nuts and chlorotic spotting in the fronds. Lower

FIGURE 5. Close up of the crown of a coconut palm *(Cocos nucifera)* in early to mid-stage of cadang-cadang disease. Nuts are fewer than normal and small and misshapened. Necrosis of newly emerged inflorescenses is also evident.

fronds become yellow-bronze and necrosis increases in the inflorescences (Figure 5).[6-8] Palms normally die 8 to 16 years after onset of symptoms. Replacement of diseased palms is time consuming, and the prognosis for replanted trees is uncertain since neither the origin of infection nor the means of natural spread is known. Eradication of diseased palms has not

Table 1
FACTORS AFFECTING AGRICULTURAL IMPACT OF PLANT VIROID DISEASES

Disease	Causal viroid[a]	Distribution[b]	Incidence[c]	Severity[d]	Controls[e]
Avocado sunblotch	ASBV	+ + +	+ +	V	AB
Chrysanthemum chlorotic mottle	ChCMV	+	+	V	AB
Chrysanthemum stunt	CSV	+ +	+	S	AB
Citrus exocortis	CEV	+ + +	+ +	V	AB
Coconut cadang-cadang	CCCV	+	+ + +	S	C
Cucumber pale fruit	CPFV	+	+	M	B
Hop stunt	HSV	+	+ +	S	AB
Potato spindle tuber	PSTV	+ + +	+ +	V	AB
Tomato apical stunt	TASV	+	+	S	B
Tomato bunchy top	TBTV	+	+	S	B
Tomato planta macho	TPMV	+	+	S	B

[a] Abbreviations used for disease and pathogen in text.
[b] Geographic range. + = localized (one country or area) to + + + = worldwide or widespread.
[c] Amount of infection where disease occurs.
[d] Severity of symptoms in economic hosts. S = severe, M = mild or moderate, and V = variable according to strain.
[e] Current or potential control strategies. A = avoidance and selection of viroid-free propagation stocks, B = control of contamination, C = rouging of infected plants. (Avoidance of seed transmitted infections assumed under A for ASBV and PSTV.)

consistently reduced incidence of new infections, but this may be partly due to problems in early recognition of infection.[6,7] In addition to coconut palm *(Cocos nucifera)*, other hosts for CCCV include African oil palm *(Elaeis guineensis)*, buri palm *(Corypha elata)*, betel nut palm *(Areca catecha)*, manila palm *(Adonidia merrilli)*, palmera palm *(Chrysalidocarpus lutescens)*, and royal palm *(Oreodoxa regia)*.[33,64]

2. *Potato Spindle Tuber Disease*

The potato spindle tuber viroid (PSTV) has been a persistent problem for potato growers in North America since the early 20th century and has gradually spread to, or has been recognized in other countries.[2,11] Exact losses from PSTV infection are hard to document since effects vary with viroid strain, with environment and with the host cultivar. In addition, potatoes are often simultaneously affected by viruses as well as viroids and accurate estimates of loss due to each component of such a disease complex are difficult to quantify.[2] Certainly, PSTV can significantly reduce yields when all factors favor disease development. Yield reductions of over 50% have been reported in experimental trials.[2,26] Such spectacular effects are less common and probably less economically significant over time than pernicious effects of relatively mild isolates where immediate losses are less, but the disease is more likely to be unrecognized and to continue over a period of time.

PSTV causes stunting, upright growth, and some leaf distortion in the foliage of potato *(Solanum tuberosum)* and reduces size and number of tubers.[2,26] Tubers are also elongated or otherwise deformed. Potato is generally considered the only natural host of PSTV, but it has been suggested that tomato bunchy top disease is also caused by PSTV.[3]

One of the indirect effects of PSTV is that it significantly interferes with rapid and expedient movement of seed stock and plants for both commercial and experimental purposes.[34] Basically, we have learned to contain PSTV, but no permanent control has been developed, and the hazard of an epidemic remains, especially if certification and sanitation procedures are not continually maintained.

3. Chrysanthemum Stunt Disease

The entire chrysanthemum industry of the U.S. was nearly destroyed by the chrysanthemum stunt viroid (CSV) in the late 1940s.[2,4,5,10] The disease spread explosively through clonally propagated, glasshouse-grown chrysanthemums *(Chrysanthemum morifolium)* before the nature of the problem was sufficiently understood and controls in the form of disease-free mother stock and sanitation were instituted.[9] The principal symptoms in chrysanthemum, the only natural host known, include stunting, premature blooming, inferior flower quality, lack of root formation in cuttings and chlorotic blotching in the foliage (Figure 2).[2,4,10] As with PSTV, the disease is contained, but still a potential hazard.

4. Chrysanthemum Chlorotic Mottle Disease

The chrysanthemum chlorotic mottle viroid (ChCMV) has posed a threat similar to CSV in the same crop. Partly because of the prior experience with CSV, it has been contained with less damage than occurred from CSV.[10] Symptoms include chlorotic mottling and stunting. Despite some similarities of symptoms in chrysanthemum, the causal viroids of CSV and ChCMV are apparently quite distinct.[25] As noted by Keese and Symons (see Chapter 1) ChCMV has yet to be fully characterized.

5. Citrus Exocortis Disease

Exact loss figures are not available for diseases caused by citrus exocortis viroid (CEV), but it has caused severe production losses in localized situations in many countries, and its presence and disease potential have had a marked effect on disease control strategies for other diseases. Because CEV is latent in most common citrus scion varieties, including sweet orange *(Citrus sinensis)*, grapefruit *(C. paradisi)*, and mandarin *(C. reticulata)*, as well as in sour orange *(C. aurantium)*[19,35] (historically the primary rootstock for propagation of citrus in new areas around the world), its widespread distribution went unrecognized. The absence of CEV in many Asian citrus cultivars (the native home of citrus)[36-38] not only allowed their extensive successful use on CEV-sensitive rootstocks such as rangpur lime *(C. limonia)* and trifoliate orange *(Poncirus trifoliata)*, but suggests that CEV did not originate there. Interestingly, CEV really was not a problem until citrus tristeza virus (CTV) began devastating large citrus producing areas, and growers began switching from sour orange to rootstocks such as rangpur lime, trifoliate orange, and various trifoliate orange hybrids.[19] The latter are tolerant or resistant to CTV, but quite susceptible to scion transmitted CEV and exocortis problems quickly surfaced. As noted earlier, citrus exocortis disease is not lethal, but causes stunting and a resulting loss in yield. Bark scaling (Figure 6) also occurs on sensitive rootstocks.[14,27,30]

As with other tree crops, replacement of a citrus exocortis-diseased tree is not rapid, and production losses accrue from the original problem, and also from the years it takes to bring a replacement tree into production. One of the indirect losses due to CEV is that it has restricted use of CEV-sensitive rootstocks for many years in situations where they would have been the best choices in the absence of CEV. Use of virus-free budwood gradually has allowed use of CEV-sensitive rootstocks, but only after indexing procedures were developed,[39] the horticultural value of nucellar bud sources had been evaluated, and the spread of CEV as a contaminant had been recognized and controlled.[40,41] The advent of shoot-tip grafting to obtain CEV-free propagation sources has finally removed limitations to rapidly developing viroid-free budwood sources.[42]

Widespread use of mechanical hedging and pruning equipment to control citrus tree size (Figure 7) has raised concerns about transmission of CEV as a contaminant,[43] but the major risk from contamination remains with nurseries and young plantings. The only known natural host for CEV is citrus, but like PSTV, it has a broad experimental host range[2] and could be present in other hosts.

FIGURE 6. Exocortis bark scaling symptoms in lemon *(Citrus limon)* tree grafted on trifoliate orange *(Poncirus trifoliata)* rootstock. Note that scaling is confined to trifoliate orange tissue even though both varieties are systemically infected. Tree with severe bark scaling are normally stunted, but stunting may also occur in the absence of marked bark scaling.

6. Avocado Sunblotch Disease

Avocado sunblotch viroid (ASBV) has posed certain problems in producing healthy avocado trees *(Persea americana)* and in the movement of propagation stocks,[44] but has not caused great agricultural losses. Its potential for damage is probably somewhat restricted by the limited cultivation of the host crop and use of viroid-free propagation materials. Fruit symptoms are illustrated in Figure 4. Seed and pollen transmission of ASBV,[45,46] however, pose some additional problems for containment of the disease and for maintaining disease-free certified propagation sources. ASBV has been transmitted to cinnamon *(Cinnamomum zeylonicum)*, camphor *(C. camphora)*, and coyo *(Persea schiediana)*.[47,48]

7. Cucumber Pale Fruit Disease

Cucumber pale fruit viroid (CPFV) can cause significant losses in infected cucumber *(Cucumis sativus)* plants,[14] but since it has been a glasshouse problem and of sporadic occurrence only in The Netherlands, its overall economic impact has not been great. Until the epidemiology of CPFV is better understood, its full disease causing potential cannot be evaluated. The close relationships discovered between CPFV, hop stunt,[3,49] and a grapevine viroid[50] suggest that movement has occurred between different host systems.

8. Hop Stunt Disease

Hop stunt viroid (HSV) has received increasing attention in Japan.[11,15,21,51] Hop stunt disease has occurred at relatively high incidence in some locations in Japan,[11] but has not

FIGURE 7. Mechanical hedgers and toppers are commonly used to prune citrus groves where citrus is grown on a large scale, and many operations are mechanized due to high labor costs. Trees are hedged and topped periodically to maintain a constant canopy dimension in the grove. This extensive pruning increases the risk of spread citrus exocortis viroid as a contaminant if infected trees are present in the planting. Equipment can be decontaminated chemically between use in different groves, but continuous decontamination to prevent tree to tree spread of CEV within a grove is difficult. Heat is not effective and chemicals which are instantly effective are generally hazardous and/or corrosive. (Photo courtesy S. Hedden.)

yet been found in other countries. Most hops in Japan are imported cultivars and infection from another crop is suspected. Diseased hop plants *(Humulus lupulus)* are stunted and cone production and quality are affected. Similarities between symptoms caused by hop stunt and cucumber pale fruit viroids have been noted,[3,24] and a close relationship has been confirmed by sequencing studies.[49]

9. Tomato Viroids

Tomato bunchy top viroid (TBTV), tomato apical stunt (TASV), and tomato planta macho (TPMV) viroids all cause severe effects on infected plants, but limited incidence restricts consideration of them as causes of major production problems. Symptoms are similar,[22] and TASV and TPMV share extensive sequence homology with each other and PSTV.[52] Bunchy top symptoms were induced in tomato by CEV,[53] but no careful comparisons have been made between CEV and TBTV. No isolates of TBTV have been available recently for further study. Again, lack of epidemiological information hampers accurate assessment of the potential problem these viroids may pose in the future.

D. New and Probable Viroid Diseases

In addition to the better characterized viroid diseases listed above, other plant diseases are now suspected of having a viroid etiology. While recent improvements in viroid purification and characterization technology have facilitated identifications of viroids, it is still often not easy to conclusively prove that a specific disease is caused by a viroid and even more difficult to determine its exact relationship to existing viroids in the absence of nucleotide sequence analysis.

Diener[2] has provided some basic criteria for establishing a viroid etiology of a disease. These basically include evidence that microorganisms or virus particles are not closely associated with the disease, that infectivity does not sediment in gravitational fields which pellet conventional viruses, that the pathogen is ribonuclease sensitive, and that its migration pattern in polyacrylamide gels indicates a molecular weight consistent for viroids (generally 80,000 to 120,000 daltons).

Several major problems often affect characterization of a disease as viroid induced. First of all, the plants in question may be affected by several pathogens. Secondly, viroids often are present only in low concentrations, even in symptomatic hosts growing under conditions favorable to the putative viroid. Thirdly, assays of infectivity may be difficult because symptoms are not well defined or are slow to develop in the diagnostic hosts available. Clarification of the viroid nature of PSTV and CEV did not occur until improved bioassays were developed. Assay procedures certainly were limiting factors in proving a viroid etiology for ASBV and CCCV, and it is doubtful that a viroid etiology would have been established for these diseases without the precedents established with PSTV and CEV.

We have not made an exhaustive attempt to locate all references to potential or possible viroid-associated diseases. From the following listing, however, it should be clear that this is an active area of investigation.

1. Apple

Koganezawa et al.[54] reported a viroid-like RNA associated with apple scar skin disease. Its presence was correlated with apple scar skin symptoms, but a causal relationship was not proven. The bioassays available may make this task somewhat difficult.

2. Burdock

Chen et al.[55] also isolated viroid-like RNAs from stunted burdock (*Arctium tomentosum* and *A. lappa*) plants.

3. Carnations

Lommel and Morris[56] reported a viroid-like RNA associated with a stunting condition in carnations, but did not confirm infectivity in the absence of carnation mottle virus which was also associated with the condition.

4. Citrus

Schlemmer et al.[57] isolated an infectious, viroid-sized RNA from citron plants showing citrus exocortis disease symptoms of variable severity which migrated more rapidly in polyacrylamide gels than CEV and differs in host range. The sequence homology between this viroid (citron variable viroid, CVaV) and the larger CEV has not yet been determined. Very recently, Duran-Vila et al.[58] found two additional viroid-like RNA species in citron plants with mild citrus exocortis disease symptoms. These are infectious in citron, but somewhat smaller (311 to 335 nucleotides) than typical CEV (371 nucleotides) and distinct from the entity isolated by Schlemmer et al.[57]

Circumstantial evidence has been presented that the cachexia (xyloporosis) disease of citrus is caused by a viroid,[59] but infectivity has not yet been associated with a viroid-sized molecule. Similarities to citrus exocortis disease include the nature of mechanical transmission, resistance to thermotherapy, and ease of eliminating the pathogen from infected stock by shoot-tip grafting.[59] Lack of a rapid bioassay has hampered efforts to better characterize the causal agent of cachexia.

5. Columnea

Owens et al.[20] isolated previously unreported viroids from *Columnea erythrophae* and several *Solanum* species. The *Solanum* viroids are probably closely related to PSTV, but the *Columnea* viroid is apparently different, based on differences in electrophoretic mobility and in homology to PSTV cDNA.[20] The *Columnea* viroid is latent in *Columnea,* but causes typical viroid disease symptoms in tomato and potato.[11,17,20]

6. Grapes

Sano et al.[50] have isolated a viroid-like RNA from five grape sources in Japan which was infectious to cucumber and showed extensive sequence homology with the cucumber isolate of HSV. Viroid-like RNAs have also been isolated from grapevines by Flores et al.[60] Infectivity in *Gynura aurantiaca* was demonstrated for one of the RNA species isolated but not for the other. A possible relationship to one or more grapevine disease of uncertain etiology was suggested, but not proven.

E. Virusoids

RNA species similar in size to viroids have been detected in association with several plant viruses, including velvet tobacco mottle virus and *Solanum nodiflorum* mottle virus.[61] Infected plants contain RNA species quite similar to viroids plus larger RNAs.[62] However, these small RNAs cannot replicate independently and do not cause disease by themselves.[63] Until an independent role is established for these viroid-like RNAs, there does not seem to be a strong argument for grouping these virusoids with viroids which do function independently. Virusoids may, however, hold clues for the origin of viroids.[63]

II. FACTORS AFFECTING DISEASE DEVELOPMENT

A. Host Susceptibility and Tolerance

Plants which do not show symptoms of infection in the presence of virus or virus-like pathogens are frequently referred to as "disease resistant". This terminology can mask an important difference in pathogen-host interaction. It is frequently important to distinguish between host susceptibility to infection and the reactivity of susceptible hosts after infection occurs. Resistant plants do not support replication of the pathogen and do not develop symptoms. Susceptible plants support replication of the pathogen but can vary in reaction from complete tolerance to those which develop striking symptoms. Tolerant hosts may be as acceptable to the grower as resistant ones in terms of crop production, but tolerant hosts do maintain a pathogen reservoir which may be undesirable for controlling the disease in other crops.

Some viroids seem to have a narrow host range and infect only a limited number of plants, even under experimental conditions (ASBV, CCCV, and ChCMV).[2,8,11] Others, such as PSTV and CEV, can infect a larger number of plants.[2] In the latter case, many of the plants infected may be relatively tolerant of viroid infection and do not show obvious symptoms. For example, Diener[2] lists over 150 suspected hosts for PSTV, but only about 20% of these show some noticeable symptoms of infection. Most commercial citrus scion varieties are susceptible to CEV infection, but do not show symptoms.[19,35] Disease occurs when CEV is present in specific grafted scion-rootstock combinations.

In some cases, the information on host susceptibility remains very incomplete. Coconut palm and several other palm species are the only hosts known for CCCV,[8,33,64] but this conclusion may be more an indication of the technical difficulties associated with experimental initiation of infection in other hosts or in detecting it once established than actual host resistance. The only bioassay currently available for CCCV is to inject palm seedlings and wait several years until symptoms appear or the viroid can be recovered and identified by PAGE or nucleic acid hybridization probes (see Section VI).[7,33,64] Similar problems apply to ASBV, but evidence for its recovery from several experimentally inoculated hosts besides avocado[47,48] suggests a more extensive host range eventually may be discovered. The host range for these and several other viroids will not be fully known until better assay and/or inoculation procedures become available. Negative results in host range tests are worthless unless infectivity of the inoculum is verified by inoculation of suitable controls.

Several plant hosts are susceptible to infection by different viroids and often show similar symptoms. Examples include PSTV, CEV, TASV, and TPMV in tomato, PSTV, and CEV in petunia *(Petunia hybrida),* potato, *Scopolia sinensis* and *Gynura aurantiaca,* and HSV and CPTV in various cucurbits.[2,3,11,24]

Marked differences in tolerance can occur between closely related plants. Different cultivars of chrysanthemum react differently to CSV and ChCMV[4,10] and differences in reactions of potato cultivars to PSTV have also been reported.[26] Some seedlings of Etrog citron *(Citrus medica)* react strongly to CEV, but others from the same source tree are not reactive even though infection can be demonstrated by indexing.

In general, there seems to be a positive correlation between symptom expression and viroid replication where hosts of differing reactivity have been inoculated with the same viroid isolate.

B. Strain Variation

Strains of several viroids have been described which differ markedly in the types of symptoms they produce.[2,3,25,31,65-67] Even though it is still not completely clear how viroid nucleic acids function to control their own replication and to modify host behavior, differences in biological effects are correlated with nucleotide differences[3,68-71] and indicate a functioning genetic system.[1]

The most extensive evidence for strain variation has been obtained for PSTV and CEV, but mild isolates of ASBV[45] and ChCMV[65] have also been observed suggesting the possibility of alternate strains of these viroids as well. Mild strain expression can easily be overlooked in the absence of a sensitive assay system,[67] and this may be one reason why different strains of some viroids have not been reported. Mild strains of PSTV were not recognized for a long time, but once discovered, were found to be more common than severe ones when appropriate assays become available.[2,67] This is not surprising in a vegetatively propagated crop such as potatoes where selection against disease is commonly practiced during propagation and severely diseased tubers would be rogued and not repropagated.

A range of isolates varying quantitatively in symptoms from very mild to very severe has been found for PSTV and CEV. Frequently, the only evidence cited for probable strain variation is symptomatology. This may be inaccurate for several reasons, but confirmation that differences in symptoms in a given host are due to strain variation alone can be difficult.

Host and climatic factors can usually be evaluated by use of appropriate controls, but the presence and effects of other viroids or viruses may be difficult to control. Since different viroids often cause somewhat similar symptoms in a given host,[11] one cannot simply assume that a difference in symptoms is due only to a strain modification. Differences in CEV symptoms have been noted in various citrus hosts and usually have been attributed to strain variation. Strains from several isolates of CEV have been purified and well characterized, but similar attempts to characterize strains from mild isolates have not been as successful because of poor yields. Recently, Schlemmer et al.[57] presented evidence that Etrog citron plants infected with a presumed isolate of CEV which produced variable, but moderate symptoms apparently are infected with a viroid of noticeably lower molecular weight. Similarly, Duran-Vila et al.[58] have found different RNA species associated with other presumed mild isolates of CEV which may not be strain variants but actually different viroids.

Caution seems advisable in interpretation of viroid strain effects until symptom differences are correlated with molecular properties. Unfortunately, the general pattern has been that mild strains have been more difficult to purify and thoroughly characterize than severe strains, and direct comparisons frequently have not been made. Continued improvements in viroid purification procedures and more sensitive assays should aid in further comparative studies.

Presence of viruses may also cause confusion about strain severity unless appropriate controls are utilized.[2] Despite the various problems noted, strain variation certainly has been documented and it is likely that as assay procedures improve, additional variants will be discovered.

The relative economic importance of mild and severe strains of viroids can be argued from several points of view. In vegetatively propagated crops such as citrus and potatoes, mild strains may actually cause greater total losses than severe strains because mild strains are less likely to be detected in foundation stocks and more likely to be widely propagated. Loss per plant is less, but the cumulative effect may be greater. In diseases where infection is not spread by vegetative propagation of the host, severity is more likely to be correlated to economic impact.

C. Environment

Symptom expression of viroid diseases is affected by climatic and environmental conditions.[2-4,11] In contrast to many plant virus diseases, viroid disease development is usually accentuated by relatively high temperature and high light intensity.[2,11,30,34,72,73] Early observations with potato spindle tuber indicated that symptoms were generally more severe in warm areas than in colder regions.[2] Subsequently, a number of studies have been done to optimize conditions for viroid disease development under experimental conditions, especially for bioassay purposes,[2,4,34,73-75] Warm conditions (28 to 32°C) favor development of CEV, PSTV, ASBV, CPFV, and HSV-induced diseases. The optimum may be slightly lower with CSV and ChCMV,[4] but adherence to the general trend seems consistent. Purification and assay studies all indicate that where high temperature conditions promote viroid disease development, viroid replication is also enhanced, indicating a direct relationship between symptom expression and viroid replication. In nearly all studies reported, high light intensity conditions have also favored viroid symptom development.[4,11,34,57,75] Symptoms also develop more rapidly and strongly under long day conditions and supplemental lighting increases symptom expression in winter months.[34,75]

Not surprisingly, viroid diseases have often been the biggest problem in crops grown under warm, high light intensity conditions. This would suggest that searches for new viroid-induced diseases or testing of host reaction to existing ones should be done under similar conditions. Some of the viroid-induced symptoms such as stunting can only be recognized when plant growth conditions are favorable for rapid growth of nonaffected plants. In the case of CEV, for example, warm, high light conditions not only enhance viroid replication, but also favor good growth of the citrus host.[72]

Singh[11] has concluded that although viroids cause more severe symptoms in potato in hot climates, they would not pose a severe threat to potato production in these areas because infected plants would be unthrifty and not perpetuated by propagation. This may be true for severe isolates and where no introduction of seed stock is made from cooler climates, but these assumptions may not always apply.

Young navel orange trees on Carrizo (*Poncirus trifoliata* × *Citrus sinensis*) rootstocks infected with severe CEV isolates showed significantly more freeze injury than trees infected with mild isolates or on the uninoculated controls.[76] Severe bark scaling similar to that shown in Figure 6 developed the year following freeze damage on many of the Carrizo rootstocks infected with the severe isolates. This scaling was more severe than normally observed on Carrizo and occurred at an abnormally early age. Since susceptibility to cold injury has not been consistently associated with CEV infection, it is possible that the effects were indirect. Cold injury in citrus is closely correlated to cambial activity and if the natural growth cycle of CEV-infected and healthy or mild strain-infected plants were out of phase,[77] a differential response to cold could be expected, but this would not necessarily be a direct effect of CEV infection.

III. TRANSMISSION OF VIROIDS

A key factor in assessing the economic significance of a plant disease is its ability to be disseminated. A disease which severely affects its target plant may not be a serious economic threat if it is not readily spread under normal growing conditions. Similarly, the ease of preventing transmission is also a factor in evaluating economic significance.

Viroids are similar to viruses in some aspects of transmission and quite different in others. The principal known means of viroid spread are via propagation, contamination (mechanical inoculation), and through pollen or seed.[2,3] These are all common to viruses as well. The major difference between transmission of viroids and viruses is that vectors such as insects and nematodes which are often highly important in dissemination of viruses, are either not major factors in the natural spread of most viroids or have not been confirmed where vector spread is suspected.[2]

A. Propagation

As the properties of viroids have become better understood, it is easy to see how these pathogens could have become major problems in crops such as avocados, citrus, chrysanthemums, hops, and potatoes, which are all commercially propagated by vegetative means. Viroid diseases affecting these crops can be transmitted to progeny generations (perpetuated) by propagation, and the potential to increase infection can be enormous. For example, 20,000 to 50,000 buds may be cut and propagated from a single citrus budwood source tree within a single year. If subpropagations from nursery trees (second cut budwood) are also made, upwards of 1 million infected trees can be propagated from a single source tree within 1 year.

The capacity for primary multiplication from a single potato or chrysanthemum plant by vegetative means is somewhat less due to the smaller number of plants which can be prepared from a single parent. In the case of chrysanthemum, however, this factor is offset by the relatively short generation time to root and increase cuttings.

Application of new tissue culture procedures for commercial propagation has also increased opportunities to propagate large numbers of plants from a single source. It may be argued that because fewer source plants are involved and rigorous indexing of each source is more feasible, the risk of disease multiplication is reduced. However, it is clear that the potential for multiplication of a disease is also increased by concentrating production on a limited base.

Aside from the sheer potential for volume multiplication of a pathogen, propagation holds other risks. Plants propagated by man are often shipped long distances to new growing locations. Within 24 hr significant amounts of inoculum can be transported halfway around the world in the form of PSTV-infected seed potatoes, CEV-infected citrus budwood, ASBV-infected avocado buds, and CSV-infected chrysanthemum cuttings. Man certainly can be given credit as a prominent long range disseminator of viroid diseases.

In both citrus and avocado, commercially propagated trees generally consist of clonally propagated scion varieties budded or grafted on seedling rootstocks. This use of a combination of germplasm sources can complicate determination of the origin of infection, especially in the case of ASBV where high rates of seed transmission may occur in the absence of symptoms in the infected seedlings. In the case of citrus, seed transmission of CEV has not been demonstrated, but most common scion varieties such as oranges, grapefruit, and mandarins are all latent carriers and thus show no visible symptoms to warn growers against their propagation on sensitive rootstocks.

B. Seed and Pollen

Seed transmission has been demonstrated for PSTV[2,11] and ASBV[45] and may occur at very low rates for CCCV,[8] but does not appear to be a major factor in commercial situations. The exception is avocado for which both seed and pollen transmission create some problems in controlling disease spread. Since true seeds are not used in commercial potato production, seed transmission of PSTV is a factor mainly in breeding and experimental studies. Seed transmission of CEV has been suggested several times, but it now appears that these cases can be more readily ascribed to unrecognized infection via contamination.[78]

C. Contamination

Several viroids can be readily transmitted mechanically by appropriate procedures and dissemination via contamination is a major factor in the spread of several viroid diseases.[2-4,11,16,21,79] Nearly 60 years ago, Goss showed that PSTV could be transmitted on contaminated knife blades and by rubbing healthy and infected seed pieces together, but the importance of mechanical transmission in disease dissemination was not fully appreciated for another 25 years.[2] Mechanical spread has now been implicated as a factor in the spread of CEV,[43] CPFV,[14] CSV,[4] PSTV,[79] HSV,[21] and TPMV.[16] In the crops affected by these pathogens, cultural practices generally create considerable opportunity for wounding and for plant to plant movement.[43,79] ChCMV is not as readily transmitted mechanically under experimental conditions as many other viroids,[2] and contamination may not be such a significant factor in its spread. ASBV has been mechanically transmitted experimentally.[80,81] No evidence has been reported for cultural spread of CCCV as a contaminant,[6,7] but it has been transmitted by pressure injection.[7,64]

Fairly extensive studies have been done on mechanical transmission of CEV,[78,82-84] and these illustrate some of the key points on mechanical transmission of viroids. Mechanical spread of CEV was unrecognized until increased use of CEV-free budwood led to repeated observations which suggested that natural spread was occurring under both field and glasshouse conditions.[40,78] The origin of these apparent natural infections remained unknown for some time and threatened extensive budwood certification programs. The eventual discovery that CEV was readily transmitted mechanically as a contaminant on knife blades[40] now seems very simple and obvious. However, at that time the properties of viroids were unknown and early attempts to transmit CEV via conventional leaf abrasion procedures had failed. A number of other possibilities were investigated before the correct simple test was finally made. Further study showed that the stem slash method and other inoculation procedures which cause vascular wounding were much more effective than leaf abrasion for transmission of CEV.[82] Similar observations have also been made for other viroids.

Other studies on CEV have indicated marked host effects on mechanical transmission at both donor and receptor levels. For example, CEV can be readily transmitted between reactive citrus hosts, but is difficult to mechanically transmit to some asymptomatic plants such as grapefruit even though these hosts are susceptible to infection by graft inoculation.[78]

In the field, accidental contamination of propagating sources and nursery plants is a major hazard which must be avoided. Extensive use of mechanical pruning devices to control tree size also presents a good opportunity to spread CEV (Figure 7).[43] Viroids generally are very heat stable and surprisingly resistant to many chemical inactivation treatments.[4,41,78,83,84] Most are quite stable in sap extracts and can survive on dry surfaces of contaminated objects for long periods.[83] Once a source of primary inoculum is present, prevention of secondary spread can be difficult.

D. Vectors

Early studies suggested the possibility of transmitting viroid diseases, especially PSTV, by a number of insects.[2,3,85] Unfortunately some of the more extensive vector tests preceded knowledge of the true nature of viroids and the ease with which they can be mechanically spread as contaminants.[2] There is little compelling direct evidence that viroids are extensively spread by vectors. However, de Bokx and Piron[86] did demonstrate transmission of PSTV in a nonpersistent manner by the aphid *Macrosiphon euphorbiae*.

The possibility of transmitting viroids by chewing insects such as grasshoppers[78,85] seems theoretically attractive; however, closer examination of feeding behavior of chewing insects suggests that the primary site of inoculation created by the first bite is in turn rapidly ingested by subsequent feeding activity. Unless the mouthparts can remain inoculative after prolonged periods of feeding, or after regurgitation occurs, most potential infections are probably quickly eliminated by the continued feeding of the possible vector.

No nematode or other soil-borne vectors for viroids have been reported. Lack of a coat protein may prevent viroids from binding to these potential vectors.

Vector movement of CCCV has long been suspected, because of the pattern of disease spread,[6] but so far, firm evidence for this has not been provided. Low rates of natural spread make it difficult to identify potential vectors circumstantially, and experimental inoculation tests with field-collected insects have been negative.[6] Some circumstances surrounding development of cucumber pale fruit disease also suggest a vector may be involved in the original infections,[14] but again, definite proof is lacking. Similarly, vectors were suggested for TBTV from circumstantial evidence.[2]

IV. VIROID INTERACTIONS

A. Strain Interactions

Several observations indicate that infection by one strain of a given viroid may affect symptom expression and/or infection by another strain of the same viroid.[2,3,25] This type of effect in virus systems has been called cross protection, and the term has been used to describe both the suppression of symptoms and the suppression of replication of the challenging agent. The distinction may not be of practical significance to growers but is important from a technical standpoint.

Preliminary evidence for strain interference among viroid strains was actually obtained before viroids were identified. Wallace and Drake[45] observed that the avocado sunblotch pathogen could be seed transmitted at high rates from symptomless trees and that the symptomless progeny seedlings were protected from challenge inoculation by symptomatic sources. Similarly, trees or portions of trees which "recovered" following an initial typical reaction to infection were also protected against challenge from severe sources. The mild and severe forms of ASBV differ in symptom expression and in seed transmission. Rates of seed transmission are very high for latent forms and generally less than 5% for severe forms.[13,45]

Fernow[67] observed that mild isolates of PSTV (PSTV-M) existed which were not readily detected by symptom expression on tomato seedlings he used as indicators. He demonstrated, however, that when tomato plants were inoculated with these mild, usually symptomless isolates and then challenged with severe PSTV isolates, some partial and temporary inhibition (cross protection) was obtained. He actually used this protection phenomenon in a practical way to detect mild isolates of PSTV.[67] Tomato seedlings were inoculated first with extracts from a plant to be indexed. If no symptoms appeared after 10 to 14 days, the plants were challenged by a severe isolate of PSTV (PSTV-S). A delay in symptom expression over that observed in challenged controls was evidence for presence of a mild isolate. Subsequently, it was found that PSTV-M is widespread. Mild and severe isolates have been characterized and differ only slightly in nucleotide sequence.[1,70,71]

Horst[65] also observed that some plants of Deep Ridge chrysanthemum, normally a reactive host for ChCMV, failed to react to inoculations. Subsequent studies showed that these plants contained an infectious agent which, when inoculated to chrysanthemum plants 8 days before challenge inoculation with ChCMV, provided complete protection against symptom devel-

FIGURE 8. Symptoms of citrus exocortis viroid (CEV) infection in Etrog citron *(Citrus medica)* indicator plants. Plant at left is infected with a severe strain, the one in the middle with a mild strain and the plant on the right is healthy. Numerous isolates of CEV, which vary in severity, can be recovered from field trees. Very mild isolates are often difficult to detect even on sensitive indicator plants. The degree of citron reaction is correlated positively with bark scaling on sensitive rootstocks, but not with tree dwarfing.

opment for 30 days.[65] In this case, protection was evaluated by symptom expression only and no attempts were reported on recovery of the challenge isolate from dually inoculated plants. Neither mild nor severe forms of ChCMV have been well characterized.

In contrast to these experiences with PSTV, ChCMV, and ASBV, there has been no evidence for strain cross protection in citrus between mild and severe isolates of CEV.[30,66] Isolates of CEV which induce only mild symptoms in Etrog citron indicators (Figure 8) have failed to protect against challenge inoculation by severe isolates. This was first reported by Kapur et al.[66] in California. A series of experiments in Florida using several sources of mild and severe CEV isolates has also failed to indicate any evidence of cross protection.[129] Citron plants were graft-inoculated with mild isolates and, after mild systemic symptoms were observed, were challenged by graft and by mechanical inoculation with severe isolates. Symptoms of the challenge isolate developed with similar intensity and incubation period in "protected" and nonprotected plants.

The lack of cross protection between CEV isolates may be due to incomplete distribution of the challenge isolate throughout supposedly systemically infected plants.[87] Although symptoms occur systemically, it is not known how completely the viroid is distributed in susceptible tissues. Alternatively, since the mild isolate of CEV used has not been sequenced or highly characterized, it is possible that the assumed mild isolates of CEV are different viroids which produce similar but milder symptoms.[57,58] Defined sequence variants of CEV[68,69] have not been evaluated for protection. Further characterization of mild and severe CEV isolates and tests of their biological interaction will be required to fully resolve this question.

B. Interaction Between Different Viroids

Niblett et al.[88] found that tomato and chrysanthemum could be used advantageously as hosts to study interactions between different viroids. They found that PSTV and CEV could be transmitted to *Chrysanthemum morifolium* cv. Bonnie Jean and would induce symptoms. They also transmitted CSV to tomato and showed that PSTV-M could be transmitted to *Gynura* where it remained symptomless. Use of these three hosts allowed biological differentiation of CEV, CSV, PSTV-S, PSTV-M, and ChCMV.[25,88] In tomato, PSTV-S and CEV produced similar severe symptoms while PSTV-M produced mild symptoms and CSV was latent; no replication of ChCMV was detected. PSTV-M protected against PSTV-S and CEV in tomato, while CSV delayed symptom expression of PSTV-S and CEV but did not prevent expression of the challenge pathogens.[25] Each viroid produced distinctive symptoms in chrysanthemum plants and CSV, PSTV-M, and PSTV-S protected against CEV. PSTV-S did not protect against ChCMV and ChCMV did not protect against PSTV-S, CEV, or CSV.[25] The authors confirmed the identities of all viroids used except ChCMV by fingerprinting ^{125}I-labeled RNA nuclease digests of each.

The diagnoses of cross protection by the authors in the preceding studies were based on host symptom expression and do not necessarily indicate that replication of the challenge pathogen was prevented. In fact, both challenge and "protecting" viroids were recovered from extracts of doubly inoculated tomatoes. A quantitative suppression of the challenge isolate may have occurred and resulted in suppression of the challenge symptoms, but this was not assayed.

Tomato may also be a host biased for revealing a limited degree of cross protection because it develops severe symptoms when young but progressively milder symptoms with age. This host effect could complement limited protection effects. In vegetatively propagated plants such as citrus and chrysanthemum, repeated flushes of growth over longer periods offer more extended opportunities for expression of the challenge isolates. However, host effects alone are not the complete explanation because Galindo et al.[16] did not observe protection in tomato between PSTV-M and tomato planta macho viroid.

Palukaitus and Zaitlin[87] have proposed an interesting model to explain cross protection, based on the concept that (+) sense RNA molecules of the protecting isolate will bind to (−) sense "template" strands of the challenge virus and prevent transcription of the challenge isolate.[87] They propose that RNA homology between viruses or virus strains need not be complete for inhibition to occur. Binding may only need to occur at a specific site such as the 3′ end. Based on this model, inhibition between viroids with considerable homology such as CSV, PSTV, and CEV could be predicted. Lack of protection does not necessarily negate this hypothesis. If the protecting strain is not completely invasive, or does not replicate well in a given host, then inhibition on the host level may not occur because the challenge isolate is not sufficiently exposed to binding by the protector. Mild isolates of CEV apparently occur in lower titer than severe isolates, but it is not clear whether this reflects less complete invasiveness of the mild isolates or lower rates of replication. However, either factor could account for the observed lack of protection.

C. Viroid-Virus Interactions

Several examples have been reported of interactions between a viroid infection and infection with an unrelated plant virus. Singh[89] reported a severe necrotic disease in Kennebec potato plants which was caused by a dual infection by PSTV and potato virus Y. Neither agent alone caused the severe disease symptoms and the combination was effective only when PSTV was the primary infecting agent. Raju and Olson[9] observed that CSV symptoms occurred in the indexing cultivar Blanche only when chrysanthemum virus B infection was present.

Presence of a latent virus in *Gynura aurantiaca* did not drastically affect the fact that expression of CEV symptoms occurred, but CEV symptoms were noticeably stronger in dually infected plants.[90] Severe isolates of citrus tristeza virus (CTV) often interfere with expression of CEV symptoms in Etrog citron indicators. Citron plants infected with CTV are often severely stunted and show an upward arc cupping of the leaves in contrast to the leaf epinasty symptoms produced in vigorous new growth by CEV (Figure 8). The interaction between CTV and CEV can be manipulated by adjusting the growing conditions. If dually infected plants are placed under high temperature conditions the effects of CTV are masked and the effects of CEV enhanced.

Other interactions between viroids and viruses probably exist and new ones will likely be discovered. This is especially probable in vegetatively propagated crops in which latent virus infections may exist unnoticed and be extensively propagated. Introduction of the second pathogen by other means is all that is required to create a reactive combination. The role of the primary latent infection may remain unrecognized. Research studies frequently focus on detection and characterization of individual pathogens and it is easy to forget that multiple infections frequently occur in the field and complicate an analysis of pathogen effects.

D. Viroid-Fungal Interactions

Rossetti et al.[91] reported that CEV infection reduced the rate of development of *Phytophthora* lesions on the citrus trunk. These effects were observed with both bark-scaling and milder, nonscaling isolates of CEV. They also occurred in the scion as well as the rootstock, so were apparently not due to localized changes in the CEV-sensitive part of the grafted combination. Conversely, it has been suggested that CEV-induced bark-scaling can make some rootstocks more susceptible to fungal infection, especially *Phytophthora* root rot in rangpur and sweet limes. These observations are not necessarily contradictory since the work by Rossetti et al. involved experimentally inoculated plants and pertained to lesion development after infection.

V. EPIDEMIOLOGY

The etiology of viroid diseases has been greatly clarified in recent years and characterization of the special properties of viroids has contributed to an improved understanding of viroid disease development. However, major aspects of the epidemiology of many viroid diseases remain virtually unknown.

For CEV, PSTV, ChCMV, CSV, and probably HSV, it is reasonable to assume that most disease situations arise from spread via propagation or contamination or a combination of both. The extent of disease development which results is governed by inoculum potential and the host and environmental factors which are pertinent to transmission and symptom expression. Similar factors apply to ASBV with some additional possibilities from seed or pollen transmission.[46] In all of the diseases above, the major inoculum reservoir is apparently within the target crop.[2] What is not clear is the origin of primary infection within each host.

Viroids may have moved into several crops from a common origin. The similarities between HSV, CPFV, and a grapevine viroid[24,49,50] could be explained by this, but natural reservoirs outside of the affected crops have not been located.[14,21] If each viroid described has arisen *de novo* in the specific crop which is the major host, a common progenitor or convergent evolution of similar structures is indicated.[52] The marked differences in size between most viroids and CCCV and ASBV and the lack of homology between ASBV and other viroids suggests some variability in origin (see Chapter 1). Resolution of this question obviously has a significant bearing on understanding viroid diseases and in evaluating different control strategies.

The inoculum reservoir for CCCV is even less clear.[6] Despite extensive research, it is uncertain if the primary inoculum reservoir for CCCV is infected palms or another host, and the means of natural spread remains unknown. A positive correlation of disease incidence to palm age and a negative correlation to latitude have been established,[6,7] but these associations have not indicated the mode of disease spread.

Distribution studies may hold more useful clues. It now appears, despite some reports to the contrary, that CCCV spreads only slowly to new areas and that it did not erupt rapidly from a single focal point.[6] Also, disease incidence within affected areas fluctuates and periodic epidemics occur at different times in different areas. The time frame for CCCV is so slow, however, that patterns do not rapidly emerge. Roguing experiments should theoretically indicate whether or not infected palms are the primary source of inoculum.[6] However, low rates of infections and, until recently, lack of procedures to recognize subclinical infections promptly have complicated this approach.

The natural infections of CCCV in buri palms indicate that pollen or seed transmission is unlikely since infected buri palms die before flowering.[6] Soil borne spread also seems unlikely. Recent discovery of latent experimental palm hosts for CCCV[64] may provide new clues.

Contamination can account for secondary spread of CPFV within glasshouse-grown cucumbers and no evidence for seed, soil borne, or aphid transmission has been reported.[14] The origin of primary infection of CPFV remains unknown, but the problem gradually seems to be spreading to glasshouses in different areas. Time of crop planting does affect disease incidence.[14]

Spread by contamination is apparently involved in movement of the tomato viroids within affected fields,[2,16] but how initial infections occur and from what sources has not been defined. Disease incidence varies from year to year in the areas affected and suggests that a dynamic system is involved.

VI. CONTROL

Control of viroid diseases is certainly a prime concern of growers who raise crops which can be devastated by them. A great deal of research also has been done in the interest of improving control. Interestingly, several of the basic strategies for controlling viroid diseases were formulated empirically nearly 50 years before the first viroid molecule was characterized.[2] As the unique properties of viroids became known, it became clearer why certain control steps were effective and why others had failed. Increasing knowledge of viroid biochemistry has also allowed development of improved detection procedures and more appropriate procedures to prevent spread by contamination.

Viroid diseases provide a graphic illustration that opportunities to control a disease effectively are created by obtaining a clear understanding of the properties and characteristics of the causal pathogen.

A. Detection

Rapid, accurate, and sensitive identification is a key component to viroid disease control strategies, but these identification parameters have not been easy to achieve for all viroids. As noted previously, severe strains may be latent in some hosts or under unfavorable conditions and mild strains may not express symptoms even in reactive hosts. Concentrations of viroids are generally low even under conditions favorable for replication and are sensitive to strain, host, and environmental effects. Furthermore, immunological and electron microscopy assay procedures so convenient for many plant viruses are not applicable.

1. Biological Tests

Until recently, indexing on diagnostic indicators has been the primary or only procedure availabe for detecting viroid infection.[2,39,67] Attention has been focused on new molecular detection procedures, but as pointed out recently by Owens and Diener,[92] biological testing remains an important tool for certain purposes. Biological assays may well remain the most sensitive and reliable means to indicate potential strains. However, there are a number of problems associated with viroid identifications which are based solely on biological testing. Many of these are implicit from information in other sections of this chapter. In some cases, no rapidly reacting diagnostic host has been found. Indexing for CCCV is especially laborious[6] and tests for ASBV[13,73] and for the dwarfing effects of CEV[29] are also very time consuming. Problems remain even for viroids such as PSTV for which rapid diagnostic indicators have been described. Mild isolates often do not produce easily recognized symptoms and even severe isolates may be overlooked if conditions are not favorable.[34,72,74] Several indicators for viroids, such as tomato, *Scopolia* sp., and *Gynura* sp., react similarly to a number of different viroids[2,3,11] and are not highly diagnostic.

2. *Gel Electrophoresis*

The extraction and gel electrophoresis procedures originally used to isolate and purify viroids have been adapted for detection of some viroids.[7,74,75,93-98] Under appropriate conditions a viroid band can be detected in polyacrylamide gels from relatively small amounts of tissue in a relatively short period of time. This PAGE procedure, however, may be laborious for screening large numbers of samples, especially if multiple steps are needed to eliminate contaminating host components. Viroid concentration is generally correlated with symptom severity and is relatively low even in highly reactive host-viroid combinations. Unfortunately, PAGE usually works best on the samples which contain viroids easiest to detect on indicators and least well on those viroids for which detection on indicators is most difficult and for which auxiliary methods are most needed. Increasing the sample load often leads to masking of the faint viroid band in the gel by other components such as host DNA, unless steps are taken to selectively enhance viroid concentration.

Despite the inherent problems, PAGE procedures have been used effectively for detecting CCCV,[7] ASBV,[94,95] and mild isolates of PSTV,[74,98] all of which are difficult to detect with existing biological assay procedures. Recent developments, including two-dimensional electrophoresis and silver staining, have increased sensitivity and reduced the need for preparative separations.[99]

PAGE procedures are also very useful for establishing preliminary evidence for viroid etiology of diseases of unknown causes.[2,50,54,55,58,60,81] If infectivity can be associated with an eluate from an area on a gel where viroids would normally be found after electrophoresis of appropriately prepared extracts, the presence of a viroid pathogen is strongly indicated even in the absence of a visible zone.

3. *Serology*

No evidence has been presented that viroid-specific proteins are produced in infected plants even though some host-encoded proteins may be selectively increased.[2,3,100] Thus, despite some early attempts, serological detection is unpromising.[2,101] While ds-RNAs can be detected serologically, antibodies to ds-RNA did not react to PSTV viroid preparations.[102] Even if reaction had occurred, low concentration of viroids in infected plants would not favor this approach. French et al.[103] did detect a circular ds-RNA in PSTV-infected tomatoes using serologically specific electron microscopy.

4. *Nucleic Acid Hybridization Analysis*

The most promising approach to rapid diagnosis of specific viroid infections in large numbers of samples seems to be use of nucleic acid hybridizaion techniques.[92] Assays may be done in solution or on a solid matrix (dot-blot). The latter is more convenient for large-scale testing. PSTV, CEV, CSV, CCCV, and ASBV have already been detected using cDNA hybridization tests,[33,68,92,104-106] and this procedure will undoubtedly be extended to others.

Several approaches have been used to obtain labeled probes. The simplest approach has been to end-label purified viroid preparations.[107] When melted, these can hybridize with the respective positive or negative sense region of the sample. The amount of probe, however, is limited to the amount of purified viroid available.

A virtually unlimited supply of labeled probe can be obtained by cloning a cDNA to the viroid RNA and labeling in vitro by nick translation.[92] Other approaches to probe development were recently reported by Bar-Joseph et al.[108] and by Barker et al.,[129] who synthesized synthetic oligonucleotides complementary to short sequences of ASBV nucleotides and used them in dot-blot procedures to identify ASBV in leaf extracts.

Once a satisfactory probe is obtained, the major problem is to develop a simple procedure to prepare samples so that sufficient viroid is present to be readily identified and to avoid nonspecific reactions.[92] Since host, viroid isolate, and environment all affect viroid titer, pertinent positive and negative controls are essential.

It is usually relatively simple to substantiate positive results. It is frequently much more difficult to prove negative results. This, however, is often critical to viroid disease control programs which involve certification of propagation stocks as being viroid-free. All the new technologies have not removed the need to optimize conditions for detection, the desirability to confirm a diagnosis by repeated testing and the need for skilled personnel with extensive experience in the techniques used.[109] Biological procedures, PAGE, and nucleic acid hybridization can all be used to complement each other and to improve confidence in negative results.[74] No "best" procedure exists for all applications.

B. Avoidance (Certification)

One of the earliest strategies to control viroid diseases was to avoid propagation of obviously infected materials. This has been used very successfully for clonally propagated crops such as potato[2,11] and chrysanthemum.[5,9] Much of the success for this strategy can be attributed to the fact that viroid diseases are generally not vectored within or between crops except by man.

Successful use of certification schemes normally requires effective means of indexing, especially when infections may be latent in some cultivars or in one of the components of a grafted plant such as citrus or avocados. Another requirement for successful employment of a certification program is ability to obtain viroid-free mother stock for propagating. If adequate indexing procedures are available, viroid-free plants may simply be located by extensive screening of existing commercial stocks.

Thermal therapy procedures widely used to free plants of virus infection have generally been ineffective against viroids, although modification in symptoms has been observed in some heat-treated materials.[110] Knowing that viroid replication is favored by high temperatures and high light intensity, Lizarraga et al.[111] used a combination of low temperature conditions and meristem culture to free potato plants of PSTV. Shoot-tip grafting (a meristem transplanting procedure) has been used successfully to free many citrus cultivars of CEV infection[42] and meristem culture was used for isolation away from HSV.[112]

In the case of ASBV and PSTV, avoidance or certification strategies must also deal with the possibility of seed transmission. This has been a persistent problem in the avocado, where infected seed source trees and progeny seedlings are often latent carriers and ASBV infection does not become apparent until infected seedlings are grafted with healthy scion.[45] Use of hybridization probes or PAGE procedures should reduce this problem.[17]

Where adequate technical support is available, certification should remain a viable, if somewhat expensive, control for diseases such as CEV, ChCMV, CSV, ASBV, HSV, and PSTV. For diseases such as CCCV, TPMV, CPFV, and TBTV, where the origin of primary infection is unclear and host plant propagation is from seed, avoidance and certification are less pertinent. Continued improvement in detection procedures, especially for mild strains or for severe strains under conditions unfavorable to their expression, will aid use of avoidance and certification schemes.

C. Sanitation and Decontamination

Spread of viroids on contaminated tools or hands during various horticultural operations (Figure 7) is a serious problem which must be addressed to control several viroid diseases.[2,4,21,43,79] Much of the spread of viroids within a crop system is due to contamination, and this may also account for introduction of some viroid diseases into a given crop from other sources (i.e., TPMV and TBTV).[2] Transmission via contamination can be significantly multiplied for CEV, CSV, HSV, and PSTV where an unrecognized initial viroid contamination of a propagating source plant can then be unwittingly increased by propagation.

The chemistry of the viroid molecule also presents some unique problems for decontamination. Heat treatment, which is highly effective against most viruses and other pathogens with pathogen-specific nucleoprotein components, is ineffectual against viroids.[4,83,84] Viroids are precipitated from solution by alcohol but remain infectious when the alcohol is removed. The secondary structure of the viroid molecule also confers surprising stability to a ''naked'' nucleic acid. Viroids can remain infectious for long periods on contaminated surfaces[83] and are not immediately inactivated by nucleases in plant saps. Detergents and commercial germicides also are not effective, at least not on brief exposure.[41,84]

Exposure time is a key factor in practical application of inactivating strategies and has given rise to some misunderstanding about what ''works'' and what does not. For example, ribonucleases can certainly inactivate viroid molecules under appropriate conditions[113] and infectivity can also be lost in strong bases,[41] but these treatments may not be practical where a high level of inactivation is needed within several seconds.[4,41]

Worker safety is a further consideration, and treatments which can instantly inactivate a viroid molecule generally entail some hazard to the worker or to the tools being decontaminated.

The most effective practical viroid decontamination treatment devised to date is probably brief exposure of contaminated surfaces to a dilute (0.25%) solution of sodium or calcium hypochorite.[4,84] This treatment is effective, and the material is inexpensive and readily available. Hypochlorite solutions are somewhat noxious to use and also are highly corrosive, but have fewer overall disadvantages than other procedures suggested. Where longer exposures can be used, ribonuclease solutions provide a safe, noncorrosive alternative.[113]

In a different approach, Singh[114] tested several materials as protectants applied to plants to be exposed to infection and found that piperonyl butoxide could reduce infection.

Because of the problems associated with eliminating viroid contamination after it has occurred, reducing the possibilities for contamination is highly desirable. Isolation of viroid-free plants from potential sources of infection is a key component in control.[5,9,11] Risk of contamination is also why viroid-tolerant cultivars may be a hazard[26] and undesirable in situations where other cultivars are intolerant and could be infected from the latent host.

D. Host Tolerance and Resistance

Development of tolerant or resistant cultivars has been a major strategy for controlling many plant diseases, especially those for which risk of natural infection is great. Only limited attention has been directed to this approach for viroid diseases. In fact, more effort has probably been spent toward finding highly reactive plants for diagnostic uses than for developing tolerant or resistant cultivars.[2,39,67,72,88,90,115] There are several reasons for this. Adequate control of some of the more important viroid diseases has been achieved by avoidance and certification schemes.[2,9] Breeding for and evaluating disease resistance is very difficult in a number of the crops affected. Finally, with several viroid diseases, there has been little pressure to develop controls because of limited distribution or incidence of disease.

Probably the viroid disease situation where development of tolerance or resistance would have the greatest practical advantage is control of CCCV. Some efforts have been made already to select or develop tolerant or resistant coconut palms.[6,7] This, however, remains a formidable task. Zelazny et al.[6] estimate that several hundred years could be required to achieve control through continuous selection for field resistance.

Experimental injection procedures to deliberately infect palms are being used for an organized screening program. To avoid delays from the long incubation period for symptom expression, PAGE assays or nucleic acid hybridization probes are being used to evaluate resistance in relatively young palms prior to the normal time required for symptoms to be expressed.[64]

As indicated previously, in crops where disease susceptible cultivars would continue to be extensively cultivated, introduction of tolerant new cultivars could be a potential hazard if they become infected. In contrast, resistant new cultivars would decrease the potential inoculum reservoir.

E. Environment

Damage from viroid diseases could be minimized in some situations by modification of environment where this is controllable (glasshouses) or by selection of growing area. For example, PSTV has been less severe in cool areas than in warm ones.[11] Environmental manipulation will obviously be of only limited use from a practical standpoint for direct control. However, modification of the environment to favor expression of viroid disease can be important indirectly for improving indexing conditions for certification work and germplasm screening.[34,73]

F. Cross Protection

Mild strain cross protection has been suggested as a control approach for some virus diseases such as TMV,[116] citrus tristeza,[117] and more recently, papaya ringspot.[118] Although interference in symptom expression has been demonstrated between some viroid strains[25] the effectiveness of cross-protection for viroid disease control is questionable. Strain interference has not been demonstrated in some viroid systems and often is temporary where it has been observed. No mild strains have been observed for some viroids, but possibly could be induced experimentally. For example, the justification for cross protection would be strongest for CCCV,[7] but mild strains of CCCV have not been described yet. Even if the potential were better, it is questionable whether cross protection should be the control strategy of choice for viroid diseases on both practical and theoretical grounds. Deliberate use of a pathogen entails some risk to the crop being protected and possibly to nontarget crops in the area. In the three virus diseases mentioned, these risks are accepted because availability of tolerant hosts is inadequate and uncontrollable natural spread of the pathogen can be anticipated. For those viroid diseases where cross protection might be effective, an uncontrollable challenge infection is unlikely, and thus the risk of potential loss from the effects of the mild isolate is less acceptable.

Palukaitus and Zaitlin[87] have made an intriguing suggestion that "self protected" plants could be developed by modifying the plant genome. This may be more feasible for viruses than viroids because of differences in replicative pathways.

G. Future Efforts

Prevention of some future viroid disease problems could be accomplished by continued efforts to identify sources of viroid infection in cultivated and noncultivated plants.[17] Dot-blot procedures could be further refined and utilized more widely for surveys by regulatory agencies to rapidly survey for known viroids in additional crops. PAGE procedures could be used to screen for new, previously unsuspected viroid infections, such as the *Columnea* viroid.[20] If found, these could be evaluated for disease potential and appropriate steps initiated to prevent a problem.

VII. VIROIDS AS CITRUS TREE DWARFING AGENTS

Viroid infections, in common with most virus infections, are generally regarded as deleterious to plant health, even though visible symptoms may be mild. For this reason, much effort has been expanded to develop propagation sources of citrus free of CEV infection and to maintain budwood source trees in disease-free condition. Thus, it is rather surprising at first glance that deliberate use of CEV has been proposed to produce dwarfed citrus trees.[119-121] There are several considerations, however, which favor this approach.

A. Economic Considerations

Labor costs for harvesting, spraying, and other grove care operations are major production expenses and can be reduced by utilizing trees with smaller, more compact canopies.[122] In many areas trees are mechanically hedged or pruned frequently to control tree canopy size.[121] Even though this pruning may reduce the total fruit yield per unit land area of grove, the net financial return can be increased by reducing production and harvesting costs.

Mechanical hedging has several disadvantages which include cost, temporary crop loss from hedged surfaces, rapid regrowth of vegetative shoots, creation of numerous pruning wounds and need for removal of debris. Growers would like to maintain a desirable tree canopy volume without frequent pruning, if possible.

As in other fruit crops, many growers have also become interested in precociously fruiting, high density plantings. Higher development and planting costs can be offset if high rates of production can be achieved early in the life of the grove.[122] Conventionally spaced citrus groves (approximately 250 trees per hectare) do not reach full production for 8 to 12+ years after planting. This means a relatively slow response to changing market demands and/or recovery from grove injuries due to weather or diseases. High density plantings of healthy, normally vigorous citrus trees can attain high production at an earlier age, but maintenance due to tree crowding becomes a problem.[122,123]

B. CEV Dwarfing Effects

CEV-infected trees propagated on susceptible rootstocks are stunted to varying degrees and also tend to fruit at an early age.[119,121,123-125] In contrast to many virus-induced stock/scion diseases, the CEV effect is apparently not confined to a narrow area at the budunion, but seems to have a generalized effect on the whole root system, possibly due to modification in growth regulator activity.[27,126] These factors have led to speculation that by planting these viroid-affected smaller trees at an appropriately closer spacing, the grower could achieve early production and have an easily maintained grove with reduced tree volume.

The degree of stunting can be varied with the selection of the CEV isolate used.[123,124] Very severe isolates would normally not be used because the stunting effects would be too extreme and tree vigor would be inadequate.

An additional factor encouraging deliberate use of CEV as a tree size control agent is that, although orange trees grafted on trifoliate orange or other sensitive rootstocks may be stunted by the proposed "dwarfing isolate", tree longevity and appearance remains good and fruit size and quality is not adversely affected (Figure 9).[119]

Regardless of the mechanism of viroid-induced stunting, the idea of dwarfing citrus trees deliberately with CEV arose from favorable grower and horticulturist observations of trees which had been accidentally or unknowingly propagated in an infected state and produced trees with desirable attributes.[27,123] In effect, customer acceptance was demonstrated before experimental work was done.

C. Characterization of Dwarfing Effects

A major problem in the practical application of viroids as citrus tree dwarfing factors is

FIGURE 9. Citrus exocortis viroid-induced dwarfing in trees propagated on CEV-sensitive rootstocks without reduced fruit quality or yield per unit volume of tree canopy. High density plantings of trees deliberately dwarfed by exocortis inoculation has been proposed in several citrus growing areas and has received special attention in Australia.

that the information obtained so far has arisen empirically from graft-propagated materials quite possibly infected with other cryptic pathogens, and the causal factor(s) have not been well characterized.[124] The most extensive trials on dwarfing factors have been conducted in Australia. Several sources of the dwarfing factors have been used, and such factors have usually been assumed to be forms of CEV.[124] One CEV isolate purified from a dwarfing factor source had a nucleotide sequence similar to "typical" CEV, while another isolate from a different source showed more extensive sequence differences.[68] Isolates from all dwarfing sources have not caused CEV symptoms in Etrog citron nor yielded a purified viroid which could be characterized as CEV.[127] Failure to readily purify and characterize a CEV viroid from some dwarfing factor sources should not necessarily be considered unusual since purification of mild isolates of CEV has generally proven difficult. It could, however, indicate that other agents are involved, such as the viroid-like species isolated from citron plants with mild exocortis symptoms by Duran-Vila et al.[58]

Bark scaling in trifoliate orange or other CEV-sensitive rootstocks is closely correlated with CEV symptom severity in Etrog citron indicators. Dwarfing, however, is not predictable from citron response.[124] Little information is available on the dwarfing effects of any highly purified CEV isolate on commercial stock/scion combinations and exact correlations of field effects with specific viroid species remain to be made. Until correlations are established between tree dwarfing and specific pathogens, doubts will remain about how to accurately and safely apply this strategy.

D. Hazards

Concern over potential hazards associated with the deliberate use of CEV or other infectious dwarfing agents has been raised. If the tree-dwarfing agents are strains of CEV of a similar viroid, the hazards of uncontrolled spread to nontarget plantings should be low, providing appropriate precautions are taken to prevent spread via contamination.[41,84] The possibility of severe viroid mutants arising from an increased reservoir of infected plants has been cited as an objection to the proposed deliberate use of mild viroid isolates. Mutation could present a problem if it occurred in a bud source tree and was widely propagated and disseminated before it was recognized. Source plants would certainly need to be evaluated periodically for changes in the viroid strain(s) they contain. Since there is little evidence for cross protection between CEV isolates, contamination of dwarfing sources with a severe isolate from another source must also be prevented. A deleterious mutation in a single tree within a commercial grove, however, would pose only a localized risk and that would be primarily to the grower who knowingly assumed it.

The natural reservoir of CEV would be increased, but the increase would be of relatively mild isolates. Millions of CEV-infected trees have been propagated on tolerant rootstacks in many countries for many years, but deliberate, planned use of a pathogen may entail potential legal liabilities to the producer that have not previously existed. For example, evidence may be required that CEV-infected fruit poses no risk to the consumer.

E. Deployment of Dwarfing Factors

Hypothetically, the ideal way to employ viroids as tree dwarfing agents would be to plant healthy trees, grow these rapidly to near the desired size and then to control further growth by inoculation with the desired, preselected viroid isolate.[125] There are several problems with this approach. Healthy, mature trees (or even partly grown trees) usually respond slowly to viroid infection,[121] even when inoculated with severe isolates, and inoculation of individual field trees has not always resulted in uniform stunting.

The easiest method of employing viroid-induced dwarfing will be to bud nursery trees with infected budwood. The problem will be to balance sufficient early tree vigor needed to get trees to a productive size and sufficient dwarfing of the mature tree to minimize the need for pruning.

A final consideration in the employment of viroids in a dwarfing strategy is that the effect is on the rootstock and not on the scion.[19,30] Rootstocks which show dwarfing effects such as trifoliate orange are not horticulturally adapted to all citrus areas and also are susceptible to other diseases which may limit their use.[128] Use of viroids for citrus tree dwarfing may eventually be used commercially in some specific situations but is unlikely to become a universal practice. An understanding of how stunting is induced by viroids may, however, suggest approaches to tree size control which would not require use of a replicating infectious agent.

REFERENCES

1. **Diener, T. O.,** The viroid-A subviral pathogen, *Am. Sci.*, 71, 481, 1983.
2. **Diener, T. O.,** *Viroids and Viroid Diseases,* John Wiley & Sons, New York, 1979.
3. **Dickson, E.,** Viroids: infectious RNA in plants, in *Nucleic Acids in Plant, Vol 2,* Hall, T. C. and Davies, J. W., Eds., CRC Press, Boca Raton, Fla., 1979, 153.
4. **Hollings, M. and Stone, O. M.,** Some properties of chrysanthemum stunt, a virus with the characteristics of an uncoated ribonucleic acid, *Ann. Appl. Biol.*, 74, 333, 1973.
5. **Lawson, R. H.,** Controlling virus diseases in major international flower and bulb crops, *Plant Dis.*, 65, 780, 1981.
6. **Zelazny, B., Randles, J. W., Boccardo, G., and Imperial, J. S.,** The viroid nature of the cadang-cadang disease of coconut palm, *Sci. Filipinas,* 2, 45, 1982.
7. **Randles, J. W.,** Coconut cadang-cadang viroid, in *Subviral Pathogens of Plants and Animals,* Maramorosch, K. and McKelvey, J. J., Eds., Academic Press, New York, 39, 1984.

8. **Randles, J. W. and Imperial, J. S.,** Coconut cadang-cadang viroid, in *CMI/AAB Descriptions of Plant Viruses,* No. 287, Commonwealth Mycological Insitute, Kew, England 1984.
9. **Raju, B. C. and Olson, J. C.,** Indexing systems for producing clean stock for disease control in commercial floriculture, *Plant Dis.,* 69, 189, 1985.
10. **Horst, R. K., Langhans, R. W., and Smith, S. H.,** Effects of chrysanthemum stunt, chlorotic mottle, aspermy and mosaic on flowering and rooting of chrysanthemums, *Phytopathology,* 67, 9, 1977.
11. **Singh, R. P.,** Viroids and their potential danger to potatoes in hot climates, *Can. Plant Dis. Surv.,* 63, 13, 1983.
12. **Schwinghamer, M. W., Scott, G. K., Mallinson, F. K., Tesoriero, L. A., and Morrison, W. L.,** Potato spindle tuber viroid: an extensive infection in the New South Wales potato breeding program, *Abstracts, 4th Int. Cong. Plant Pathol.,* Melbourne, 1983, 122.
13. **Dale, J. L., Symons, R. H., and Allen, R. N.,** Avocado sunblotch viroid, in *CMI/AAB Descriptions of Plant Viruses,* No. 254, Commonwealth Mycological Institute, Kew, England, 1982.
14. **Van Dorst, H. J. M. and Peters, D.,** Some biological observations on pale fruit, a viroid-incited disease of cucumber, *Neth. J. Plant Pathol.,* 80, 85, 1974.
15. **Momma, T. and Takahaski, T.,** Development morphology of hop stunt viroid-infected hop plants and analysis of their cone yield, *Phytopathol. Z.,* 110, 1, 1984.
16. **Galindo, A. J., Smith, D. R., and Diener, T. O.,** Etiology of Planta Macho, a viroid disease of tomato, *Phytopathology,* 72, 49, 1982.
17. **Diener, T. O.,** Subviral pathogens: viroids and prions, *Plant Dis.,* 68, 4, 1984.
18. **Singh, R. P.,** Experimental host range of the potato spindle tuber 'virus', *Am. Potato J.,* 50, 111, 1973.
19. **Wallace, J. M.,** Virus and viruslike diseases, in *The Citrus Industry,* Vol. 4, Reuther, W., Calavan, E. C., and Carman, G. E., Eds., University of California, Berkeley, 1978, 67.
20. **Owens, R. A., Smith, D. R., and Diener, T. O.,** Measurement of viroid sequence homology by hybridization with complementary DNA prepared *in vitro, Virology,* 89, 388, 1978.
21. **Yaguchi, S. and Takahaski, T.,** Survival of hop stunt viroid in the hop garden, *Phytopathol. Z.,* 109, 32, 1984.
22. **Diener, T. O.,** Viroids and their interactions with host cells, *Ann. Rev. Microbiol.,* 36, 239, 1982.
23. **Singh, R. P. and Clark, M. C.,** Similarity of host response to both potato spindle tuber and citrus exocortis diseases, *FAO Plant Protect. Bull.,* 21, 121, 1973.
24. **Sano, T., Sasaki, M., and Shikata, E.,** Comparative studies on hop stunt viroid, cucumber pale fruit viroid and potato spindle tuber viroid, *Ann. Phytopathol. Soc. Jpn.,* 47, 599, 1981.
25. **Niblett, C. L., Dickson, E., Fernow, K. H., Horst, R. K., and Zaitlin, M.,** Cross protection among four viroids, *Virology,* 91, 198, 1978.
26. **Pfannenstiel, M. A. and Slack, S. A.,** Response of potato cultivars to infection by the potato spindle tuber viroid, *Phytopathology,* 70, 922, 1980.
27. **Benton, R. J., Bowman, F. T., Fraser, L., and Kebby, R. G.,** Stunting and scaly butt of citrus, *N. S. W. Dept. Agric. Sci. Bull.,* 70, 1950.
28. **Cohen, M., Castle, W. S., Phillips, R. L., and Gonsalves, D.,** Effect of exocortis viroid on citrus tree size and yield in Florida, in *Proc. 8th Conf. Int. Org. Citrus Virol.,* Calavan, E. C., Garnsey, S. M., and Timmer, L. W., Eds., IOCV, University of California, Riverside, 1980.
29. **Fraser, L. R. and Broadbent, P.,** Variation in symptom expression of exocortis and gummy pitting in citrus trees on *Poncirus trifoliata* rootstock, in *Proc. 8th Conf. Int. Org. Citrus Virol.,* Calavan, E. C., Garnsey, S. M., and Timmer, L. W., Eds., IOCV, University of California, Riverside, 1980, 201.
30. **Semancik, J. S.,** Citrus exocortis viroid, in *CMI/AAB Descriptions of Plant Viruses,* No. 226, Commonwealth Mycological Institute, Kew, England, 1980.
31. **Singh, R. P.,** Occurrence, symptoms, and diagnostic hosts of strains of potato spindle tuber virus, *Phytopathology,* 60, 1314, 1970.
32. **Boccardo, G., Beaver, R. G., Randles, J. W., and Imperial, J. S.,** Tinangaja and bristle top, coconut diseases of uncertain etiology in Guam and their relationship to cadang-cadang disease of coconut in the Philippines, *Phytopathology,* 71, 1104, 1981.
33. **Randles, J. W., Boccardo, G., and Imperial J. S.,** Detection of cadang-cadang associated RNA in African oil palm and buri palm, *Phytopathology,* 70, 185, 1980.
34. **Grasmick, M. E. and Slack, S. A.,** Symptom expression enhanced and low concentrations of potato spindle tuber viroid amplified in tomato with high light intensity and temperature, *Plant Dis.,* 69, 49, 1985.
35. **Klotz, L. J.,** *Color Handbook of Citrus Diseases,* University of California Division of Agricultural Science, Berkeley, 1973, 67.
36. **Ling, P.,** A report on exocortis virus in Taiwan, in *Proc. 5th Conf. Int. Org. Citrus Virol.,* Price, W. C., Ed., University of Florida Press, Gainesville, 1972, 102.
37. **Miyakawa, T. and Yamaguchi, A.,** *Citrus Diseases in Japan,* Japan Plant Protection Association, Tokyo, 1981.

38. **Zhao, X. and Jiang, Y.,** The present status of citrus virus and virus like diseases in China, *Fruit Var. J.*, 39, 30, 1985.
39. **Calavan, E. C., Frolich, E. F., Carpenter, J. B., Roistacher, C. N., and Christiansen, D. W.,** Rapid indexing for exocortis of citrus, *Phytopathology*, 54, 1359, 1964.
40. **Garnsey, S. M. and Jones, J. W.,** Mechanical transmission of exocortis virus with contaminated budding tools, *Plant Dis. Rep.*, 51, 410, 1967.
41. **Garnsey, S. M. and Whidden, R.,** Decontamination treatments to reduce the spread of citrus exocortis virus (CEV) by contaminated tools, *Proc. Fla. State Hort. Soc.*, 84, 63, 1971.
42. **Navarro, L.,** Citrus shoot-tip grafting in vitro (STG) and its applications: a review, *1981 Proc. Int. Soc. Citriculture*, 1, 452, Tokyo, 1981.
43. **Wutscher, H. K. and Shull, A. V.,** Machine hedging of citrus trees and transmission of exocortis and xyloporosis viruses, *Plant Dis. Rep.*, 59, 368, 1975.
44. **Allen, R. N. and Dale, J. L.,** Application of rapid biochemical methods for detecting avocado sunblotch disease, *Ann. Appl. Biol.*, 98, 451, 1981.
45. **Wallace, J. M. and Drake, R. J.,** A high rate of seed transmission of avocado sun-blotch virus from symptomless trees and the origin of such trees, *Phytopathology*, 52, 237, 1962.
46. **Desjardins, P. R., Drake, R. J., Atkins, E. L., and Bergh, B. O.,** Pollen transmission of avocado sunblotch virus experimentally demonstrated, *Calif. Agric.*, 33, 14, 1979.
47. **Da Graca, J. V. and Van Vuuren, S. P.,** Transmission of avocado sunblotch disease to cinnamon, *Plant Dis.*, 64, 475, 1980.
48. **Da Graca, J. V. and Van Vuuren, S. P.,** Host range studies on avocado sunblotch, *S. Afr. Avocado Grower's Assoc. Yearbook* 4, 80, 1981.
49. **Sano, T., Uyeda, I., Shikata, E., Ohno, T., and Okada, Y.,** Nucleotide sequence of cucumber pale fruit viroid: homology to hop stunt viroid, *Nucleic Acids Res.*, 12, 3427, 1984.
50. **Sano, T., Uyeda, I., Shikata, E., Meshi, T., Ohno, T., and Okada, Y.,** A viroid like RNA isolated from grapevine has high sequence homology with hop stunt viroid, *J. Gen. Virol.*, 66, 333.
51. **Takahashi, T.,** Evidence for viroid etiology of hop stunt disease, *Phytopathol. Z.*, 100, 193, 1981.
52. **Kiefer, M. C., Owens, R. A., and Diener, T. O.,** Structural similarities between viroids and transposable genetic elements, *Proc. Natl. Acad. Sci., U.S.A.*, 80, 6234, 1983.
53. **Semancik, J. S. and Weathers, L. G.,** Pathogenic 10s RNA from exocortis disease recovered from tomato bunchy-top plants similar to potato spindle tuber virus infection, *Virology*, 49, 622, 1972.
54. **Koganezawa, H., Yanase, H., and Sakuma, T.,** Viroid-like RNA associated with apple scar skin (or dapple apple) disease, *Acta Hortic.*, 130, 193, 1983.
55. **Chen, W., Tien, P., Zhu, Y. X., and Tiu, Y.,** Viroid-like RNAs associated with burdock stunt disease, *J. Gen. Virol.*, 64, 409, 1983.
56. **Lommel, S. A. and Morris, T. J.,** A viroid-like RNA associated with stunting syndrome in carnations, *Phytopathology*, 73, 791, 1983.
57. **Schlemmer, A., Roistacher, C. N., and Semancik, J. S.,** A unique infectious RNA associated with citron showing symptoms typical of citrus exocortis disease, *Phytopathology*, 85, 946, 1985.
58. **Duran-Vila, N., Flores, R., and Semancik, J. S.,** Characterization of viroid-like RNAs associated with the citrus exocortis syndrome, *Virology*, 150, 75, 1986.
59. **Roistacher, C. N.,** Cachexia disease: virus or viroid, *Citrograph*, 68, 111, 1983.
60. **Flores, R., Duran-Vila, N., Pallas, V., and Semancik, J. S.,** Detection of viroid and viroidlike RNAs from grapevine, *J. Gen. Virol.*, 66, 2095, 1985.
61. **Randles, J. W., Davies, C., Hatta, T., Gould, A. R., and Francki, R. I. B.,** Studies on encapsidated viroidlike RNA. I. Characterization of velvet tobacco mottle virus, *Virology*, 108, 111, 1981.
62. **Gould, A. R.,** Studies on encapsidated viroidlike RNA. II. Purification and characterization of a viroidlike RNA associated with velvet tobacco mottle virus (VTMoV), *Virology*, 108, 123, 1981.
63. **Gould, A. R., Francki, R. I. B., and Randles, J. W.,** Studies on encapsidated viroidlike RNA. IV. Requirement for infectivity and specificity of RNA components from velvet tobacco mottle virus, *Virology*, 110, 420, 1981.
64. **Imperial, J. S., Bautista, R. M., and Randles, J. W.,** Transmission of the coconut cadang-cadang viroid to six species of palm by inoculation with nucleic acid extracts, *Plant Pathol.*, 34, 391, 1985.
65. **Horst, R. K.,** Detection of a latent infectious agent that protects against infection by chrysanthemum chlorotic mottle viroid, *Phytopathology*, 65, 1000, 1975.
66. **Kapur, S. P., Weathers, L. G., and Calavan, E. C.,** Studies on stains of exocortis virus in citron and *Gynura aurantiaca* P., in *Proc. 6th Conf. Int. Org. Citrus Virol.*, Weathers, L. G. and Cohen, M., Eds., University of California Division of Agricultural Science, Richmond, 1974, 105.
67. **Fernow, K. H.,** Tomato as a test plant for detecting mild stains of potato spindle tuber virus, *Phytopathology*, 57, 1347, 1967.
68. **Visvader, J. E. and Symons, R. H.,** Comparative sequence and structure of different isolates of citrus exocortis viroid, *Virology*, 130, 232, 1983.

69. **Visvader, J. E. and Symons, R. H.,** Eleven new sequence variants of citrus exocortis viroid and the correlation of sequence with pathogenicity, *Nucleic Acids Res.*, 13, 2907, 1985.
70. **Gross, H. J., Liebel, U., Alberty, H., Krupp, G., Domdey, H., Ramm, K., and Sanger, H. L.,** A severe and a mild potato spindle tuber viroid isolate differ in three nucleotide exchanges only, *Biosci. Rep.*, 1, 235, 1981.
71. **Dickson, E., Robertson, H. D., Niblett, C. L., Horst, R. K., and Zaitlin, M.,** Minor differences between nucleotide sequences of mild and severe strains of potato spindle tuber viroid, *Nature (London)*, 277, 60, 1979.
72. **Roistacher, C. N., Calavan, E. C., Blue, R. L., Navarro, L., and Gonzales, R.,** A new more sensitive citron indicator for detection of mild isolates of citrus exocortis viroid (CEV), *Plant Dis. Rep.*, 61, 135, 1977.
73. **Da Graca, J. V. and Van Vuuren, S. P.,** Use of high temperature to increase the rate of avocado sunblotch symptoms development in indicator seedlings, *Plant Dis.*, 65, 46, 1981.
74. **Van Gelder, W. M. J. and Treur, A.,** Testing of imported potato genotypes for potato spindle tuber viroid with a tomato intermediate/electrophoreses combined method, *EPPO Bull.*, 12, 297, 1982.
75. **Watermeyer, S. R.,** Detection of chrysanthemum stunt viroid in South Africa by polyacrylamide gel electrophoresis and bioassay, *Plant Dis.*, 68, 485, 1984.
76. **Garnsey, S. M.,** Increased freeze damage associated with exocortis infection in navel oranges on carrizo citrange rootstock, *Proc. Fla. State Hort. Soc.*, 95, 3, 1982.
77. **Rodriguez, J. L., Garcia-Martinez, J. L., and Flores, R.,** The relationship between plant growth substance content and infection of *Gynura aurantiaca* DC by citrus exocortis viroid, *Physiol. Plant Pathol.*, 13, 355, 1978.
78. **Garnsey, S. M. and Weathers, L. G.,** Factors affecting mechanical spread of exocortis virus, in *Proc. 5th Conf. Int. Org. Citrus Virol.*, Price, W. C., Ed., University of Florida Press, Gainesville, 1972, 105.
79. **Manzer, F. E. and Merriam, D.,** Field transmission of the potato spindle tuber virus and virus X by cultivating and hilling equipment, *Am. Potato J.*, 38, 347, 1961.
80. **Allen, R. N., Palukaitus, P., and Symons, R. H.,** Purified avocado sunblotch viroid causes disease in avocado seedlings, *Aust. Plant Pathol.*, 10, 31, 1981.
81. **Desjardins, P. R., Drake, R. J., and Swiecki, S. A.,** Infectivity studies of avocado sunblotch disease causal agent, possibly a viroid rather than a virus, *Plant Dis.*, 64, 313, 1980.
82. **Garnsey, S. M. and Whidden, R.,** Efficiency of mechanical inoculation procedures for citrus exocortis virus, *Plant Dis. Rep.*, 57, 885, 1973.
83. **Allen, R. M.,** Survival time of exocortis virus of citrus on contaminated knife blades, *Plant Dis. Rep.*, 52, 935, 1968.
84. **Roistacher, C. N., Calavan, E. C., and Blue, R. L.,** Citrus exocortis virus — chemical inactivation on tools, tolerance to heat and separation of isolates, *Plant Dis. Rep.*, 53, 333, 1969.
85. **Goss, R. W.,** Transmission of potato spindle tuber by grasshoppers (Locustidae), *Phytopathology*, 18, 445, 1928.
86. **de Bokx, J. A. and Piron, P. G. M.,** Transmission of potato spindle tuber viroid by aphids, *Neth. J. Plant Pathol.*, 87, 31, 1981.
87. **Palukaitis, P. and Zaitlin, M.,** A model to explain the ''crossprotection'' Phenomenon shown by plant viruses and viroids, in *Plant-Microbe Interactions: Molecular and Genetic Perspectives*, Vol. 1, Kosuge, T. and Nester, E. W., Eds., Macmillan, New York, 1984, 420.
88. **Niblett, C. L., Dickson, E., Horst, R. K., and Romaine, C. P.,** Additional hosts and an efficient purification procedure for four viroids, *Phytopathology*, 70, 610, 1980.
89. **Singh, R. P.,** A unique interaction of spindle tuber viroid and virus Y in potatoes, *Phytopathology*, 72, 962, 1982.
90. **Weathers, L. G. and Greer, F. C., Jr.,** Gynura as a host for exocortis virus of citrus, in *Proc. 5th Conf. Int. Org. Citrus Virol.*, Price, W. C., Ed., University of Florida Press, Gainesville, 1972, 95.
91. **Rossetti, V., Pompeu, J., Jr., Rodriguez, O., Vechiato, M. H., Da Veiga, M. L., Oliveira, D. A., and Sobrinho, J. T.,** Reaction of exocortis infected and healthy trees to experimental Phytophthora inoculations, in *Proc. 8th Conf. Int. Org. Citrus Virol.*, Calavan, E. C., Garnsey, S. M., and Timmer, L. W., Eds., IOCV, Univ. of California, Riverside.
92. **Owens, R. A. and Diener, T. O.,** Spot hybridization for detection of viroids and viruses, in *Methods in Virology*, Vol. 7, Maramorosch, K. and Koprowski, H., Eds., Academic Press, New York, 1984, chap. 6.
93. **Baksh, N., Lee, R. F., and Garnsey, S. M.,** Detection of citrus exocortis viroid by polyacrylamide gel electrophoresis, in *Proc. 9th Conf. Int. Org. Citrus Virol.*, Garnsey, S. M., Timmer, L. W., and Dodds, J. A., Eds., IOCV, University of California, Riverside, 1985, 343.
94. **Da Graca, J. V. and Mason, T. E.,** Detection of avocado sunblotch viroid in flower buds by polyacrylamide gel electrophoresis, *Phytopathol. Z.*, 108, 262, 1983.

95. **Utermohlen, J. G. and Ohr, H. D.,** A polyacrylamide gel electrophoresis index method for avocado sunblotch, *Plant Dis.*, 65, 800, 1981.
96. **Singh, R. P.,** Evaluation of procedures for the detection of potato spindle tuber viroid by polyacrylamide gel electrophoresis, *Can. Plant Dis. Surv.*, 62, 41, 1982.
97. **Horst, R. K. and Kawamoto, S. O.,** Use of polyacrylamide gel electrophoresis for detection of chrysanthemum stunt viroid in infected tissues, *Plant Dis.*, 64, 186, 1980.
98. **Pfannenstiel, M. A., Slack, S. A., and Jane, L. C.,** Detection of potato spindle tuber viroid in field-grown potatoes by an improved electrophoretic assay, *Phytopathology*, 70, 1015, 1980.
99. **Schumacher, J., Randles, J. W., and Riesner, D.,** A two dimensional electrophoretic technique for the detection of circular viroids and virusoids, *Anal. Biochem.*, 135, 288, 1983.
100. **Conejero, V., Picazo, I., and Segado, P.,** Citrus exocortis viroid (CEV): protein alterations in different hosts following viroid infection, *Virology*, 97, 454, 1979.
101. **Bagnall, R. H.,** Serology of the potato spindle tuber virus, *Phytopathology*, 57, 531, 1967.
102. **Stollar, B. D. and Diener, T. O.,** Potao spindle tuber viroid. V. Failure of immunological tests to disclose double-stranded RNA or RNA-DNA hybrids, *Virology*, 46, 168, 1971.
103. **French, R. C., Price, M. A., and Derrick, K. S.,** Circular double-stranded RNA in potato spindle tuber viroid-infected tomatoes, *Nature (London)*, 295, 259, 1982.
104. **Mohamed, N. A. and Imperial, J. S.,** Detection and concentration of coconut cadang-cadang viroid in coconut leaf extracts, *Phytopathology*, 74, 165, 1984.
105. **Palukaitus, P., Rakowski, A. G., Alexander, D. McE., and Symons, R. H.,** Rapid indexing of the sunblotch disease of avocados using a complementary DNA probe to avocado sunblotch viroid, *Ann. Appl. Biol.*, 98, 439, 1981.
106. **Palukaitus, P. and Symons, R. H.,** Hybridization analysis of chrysanthemum stunt viroid with complementary DNA and the quantitation of viroid RNA sequences in extracts of infected plants, *Virology*, 98, 238, 1979.
107. **Rosner, A., Spiegel, S., Alper, M., Bar-Joseph, M.,** Detection of avocado sunblotch viroid (ASBV) by dot-spot self-hybridization with a [^{32}P]-labeled ASBV-RNA, *Plant Mol, Biol.*, 2, 15, 1983.
108. **Bar-Joseph, M., Segev, D., Twizer, S., and Rosner, A.,** Detection of avocado sunblotch viroid by hybridization with synthetic oligonucleotide probes, *J. Virol. Methods*, 10, 69, 1985.
109. **Spiegel, S., Alper, M., and Allen, R. N.,** Evaluation of biochemical methods for the diagnosis of the avocado sunblotch viroid in Israel, *Phytoparasitica*, 12, 37, 1984.
110. **Stace-Smith, R. and Mellor, F. C.,** Eradication of potato spindle tuber virus by thermotherapy and auxiliary bud culture, *Phytopathology*, 70, 1857, 1970.
111. **Lizarraga, R. E., Salazar, L. F., Roca, W. M., and Schild-Rentschler, L.,** Elimination of potato spindle tuber viroid by low temperature and meristem culture, *Phytopathology*, 70, 754, 1980.
112. **Momma, T. and Takahaski, T.,** Cytopathology of short apical meristem of hop plants infected with hop stunt viroid, *Phytopathol. Z.*, 106, 272, 1983.
113. **Garnsey, S. M. and Whidden, R.,** Effects of RNase and UV light on transmission of citrus exocortis virus (CEV) by contaminated knife blades, *Proc. Am. Phytopathol. Soc.*, 1, 50, 1974.
114. **Singh, R. P.,** Piperonyl butoxide as a protectant against potato spindle tuber viroid infection, *Phytopathology*, 67, 933, 1977.
115. **Yaguchi, S. and Takahaski, T.,** Response of cucumber cultivars and other cucurbitaceous species to infection by hop stunt viroid, *Phytopathol. Z.*, 109, 21, 1984.
116. **Rast, A. Th. B.,** MII-16, An artificial symptomless mutant of tobacco mosaic virus for seedling inoculations of tomato crops, *Neth. J. Plant Pathol.*, 78, 110, 1972.
117. **Costa, A. S. and Muller, G. W.,** Tristeza control by cross protection: a U.S.-Brazil cooperative success, *Plant Dis.*, 64, 538, 1980.
118. **Yeh, S. -D. and Gonsalves, D.,** Cross protection as a possible means for control of papaya ringspot virus, *Phytopathology*, 74, 1086, 1984.
119. **Cohen, M.,** Exocortis virus as a possible factor in producing dwarf citrus trees, *Proc. Fla. State Hort. Soc.*, 81, 115, 1968.
120. **Long, J. K., Fraser, L. R., and Cox, J. E.,** Possible value of close-planted virus-dwarfed orange trees, in *Proc. 5th Conf. Int. Org. Citrus Virol.*, Price, W. C., Ed., University of Florida Press, Gainesville, 1972, 262.
121. **Long, K., Fraser, L., Broadbent, P., and Duncan, J.,** Dwarfing of citrus trees by inoculation with a pathogen, *1977 Proc. Int. Soc. Citricult.*, 3, 866, 1977.
122. **Cary, P. R.,** New concepts in citrus tree spacing and pruning practices, *1977 Proc. Int. Soc. Citricult.*, 1, 162, 1977.
123. **Bevington, K. B. and Bacon, P. E.,** Effect of rootstocks on the response of navel orange trees to dwarfing inoculations, *1977 Proc. Int. Soc. Citricult.*, 2, 567, 1977.
124. **Fraser, L. R. and Broadbent, P.,** *Virus and Related Diseases of Citrus in New South Wales,* Department of Agriculture, New South Wales, Surrey Beatty & Sons, Chipping Norton, NSW. 1979, 16.

125. **Stannard, M. C., Evans, J. C., and Long, J. K.,** Effect of transmission of exocortis dwarfing factors into Washington navel orange trees, *Aust. J. Exp. Agric. Anim. Husb.*, 15, 136, 1975.
126. **Duran-Vila, N. and Semancik, J. S.,** Effects of exogenous auxins on tomato tissue infected with the citrus exocortis viroid, *Phytopathology*, 72, 777, 1981.
127. **Schwinghamer, M. W.,** Viroids associated with graft-transmissible dwarfing of orange trees, *Abstracts, 4th Int. Cong. Plant Pathol.*, Melbourne, 1983, 28.
128. **Cohen, M.,** Blight (young tree decline) and exocortis diseases in a rootstock experiment for marsh grapefruit in Florida, *Plant Dis. Rep.*, 61, 717, 1977.
129. **Garnsey, S. M.,** Unpublished data.
130. **Barker, J. M., McInnes, J. L., Murphy, P. J., and Symons, R. H.,** Dot-blot procedure with [^{32}P] DNA probes for the sensitive detection of avocado sunblotch and other viroids in plants. *J. Virol. Meth.*, 10, 87, 1985.

Chapter 5

THE SCRAPIE AGENT

R. F. Marsh

TABLE OF CONTENTS

I. INTRODUCTION

Scrapie is a naturally occurring slow virus disease of sheep and goats which was first reported in France 250 years ago. Since that time, the infection has spread to most parts of the world and similar diseases have been recognized in man, mink, deer, and elk. These diseases have a long incubation period followed by a progressive course of neurologic dysfunction resulting in complete debilitation and death. They are produced by transmissible agents having properties which are so unusual they have been called ''unconventional viruses''.[1] This article will review these unusual characteristics and discuss the possibility that these neuropathogens may contain a viroid-like nucleic acid.

II. TISSUE ASSOCIATIONS AND SIZE

Upon reviewing studies characterizing the physicochemical properties of the scrapie agent, it soon becomes evident that the principal problem is one of purification. Most studies use either crude brain homogenates or preparations enriched for cell membranes, a subcellular fraction containing high infectivity.[2,3] Attempts to solubilize the transmissible agent using various combinations of nonionic and anionic detergents[4] or organic solvents[5] have been unsuccessful. Presently, it would seem that the agent exists as a highly cell associated, insoluble membranous or submembranous complex. With this in mind, it is not surprising that attempts to size the agent by filtration,[6] gel electrophoresis,[7,8] or rate density sedimentation[9] have produced inconsistent findings, the latter procedure yielding values of 40S to 400S. Other studies, on radiation target size,[10,11] have been more consistent, indicating an infectious agent containing a nucleic acid component of only 1×10^5 daltons, or none at all.[10]

III. EVIDENCE AGAINST A NUCLEIC ACID COMPONENT

Additional evidence that the scrapie agent may contain no nucleic acid includes studies showing that infectivity is highly resistant to UV irradiation, either alone[12] or in the presence of psoralens,[13] and failure to consistently inactivate the agent with nucleases. In contrast, infectivity is inactivated by phenol extraction[14] and by protease digestion.[15-17] These findings, together with unsuccessful attempts to detect a scrapie-specific nucleic acid,[18,19] have lead to speculation that the scrapie agent is an infectious protein.[20]

IV. POSSIBILITY OF A NUCLEOPROTEIN COMPLEX

Although viroid diseases and scrapie have certain similarities,[21] the failure to demonstrate an infectious, phenol-extracted, scrapie nucleic acid indicates that the agent is not a classical viroid.[14,22] However, it is quite possible that the scrapie agent may be composed of a small, viroid-like nucleic acid bound noncovalently to host proteins giving the agent the properties of a nucleoprotein complex (Table 1).

A. Resistance to Irradiation

The resistance of scrapie infectivity to UV irradiation can be explained by the close association of a small, essential nucleic acid with cell components. Studies characterizing lipids in membranes from scrapie-infected hamster brain have detected a UV absorbing neutral lipid which remains in high concentration in purified tissue preparations after detergent extraction.[23] Furthermore, small ribonucleoprotein (RNP) complexes found in the nuclei and cytoplasm of eukaryotic cells[24] and produced in some virus infections[25,26] are highly UV resistant,[25] some having radiation target sizes similar to the scrapie agent.[27]

The resistance of scrapie infectivity to psoralen photoadduct formation can also be ex-

Table 1
PROPERTIES OF SCRAPIE INFECTIVITY AND THEIR IMPLICATIONS FOR THE POSSIBLE NATURE OF THE SCRAPIE AGENT

Characteristic	Conventional virus	Protein particle ("prion")[20,43]	Nucleoprotein containing a viroid-like nucleic acid
Resistant to UV irradiation[10]	No	Yes	Yes
Resistant to nuclease digestion[5]	Yes	Yes	Yes
Resistant to psoralen photoadduct formation[13]	Not likely	Yes	Yes
Susceptible to protease digestion[15-17]	Possibly	Yes	Yes
Inactivated by phenol extraction[14,22]	Possibly	Yes	Yes
Susceptible to UV irradiation in the presence of chlorpromazine[31]	Yes	Not likely	Yes
Radiation target size of 1×10^5 daltons[10-12]	No	Possibly	Yes
Smallest infectious particle is 40S or less[7-9]	No	Yes	Yes
Viral particles not seen in purified preparations[33,40,41,45]	No	Yes	Yes
Self-replicating	Yes	Not likely	Yes

plained by properties of a small nucleoprotein complex. Psoralens form adducts between opposed nucleic acids.[28] An essential nucleic acid with little double strand character would resist inactivation by psoralen cross-linking. Studies using chlorpromazine, a phenthiazine antipsychotic which reacts faster with single strand than double stranded nucleic acids[29] and is capable of penetrating cellular components more effectively than psoralens,[30] have shown increased inactivation of scrapie infectivity in the presence of UV irradiation indicating that the target is an essential nucleic acid.[31]

B. Resistance to Nucleases

The most convincing evidence that the scrapie agent contains an essential nucleic acid would be inactivation of infectivity by nucleases. This may sound simple, but in practice presents formidable problems. The only means of detecting the scrapie agent is by inoculating susceptible animals and waiting for them to develop the disease. Infectivity can then be quantitated using methods of endpoint dilution or incubation interval assay,[32] neither of which are accurate within one $\log_{10}$. Therefore, it is not possible to detect a 90% effect with any degree of confidence. These difficulties are compounded by the heterogeneity of the agent's tissue associations, by its tendency to aggregate,[33,34] and by the possibility that the scrapie agent may be a nucleoprotein complex which is more resistant to nuclease degradation.[35]

C. Replication, Pathogenesis, and Cell Injury

A small scrapie nucleic acid could not code for specific replicases, therefore, would need to use cellular enzymes and, perhaps, adopt a replicative strategy similar to viroids.[36] Alternatively, the nucleic acid could be replicated in the host's small nuclear RNA pool. The latter possibility implies rapid replication, whereas scrapie infectivity slowly accumulates in brain at a rate of about one $\log_{10}$ a week for the hamster-adapted agent.[37] This does not necessarily preclude rapid replication of the scrapie nucleic acid since the infectious complex may be assembled in the cytoplasm after transport from the nucleus. One other factor which must be considered is the ability of the scrapie agent to replicate to highest titer, or perhaps exclusively, in nerve cells. This seems to suggest a narrow requirement for replicative conditions, although it is possible that the nucleic acid may be replicated in a wide variety of cells but only forms infectious complexes in nerve cells.

One of the features of unconventional viral diseases is the lack of evidence for activation of any host defense mechanisms. There is no inflammation, no specific antibody produced, and no interferon production. This would be consistent with a nucleoprotein model for the scrapie agent in which a low molecular weight nucleic acid, too small to code for specific proteins, is able to combine with host proteins forming a complex which would likely stimulate little host response. Small RNA leaders of vesicular stomatitis virus[38] and adenovirus[39] have been shown to associate with host proteins producing small RNP particles capable of causing significant biologic effects.[25] If a small scrapie nucleic acid is able to bind to host proteins, it could interfere with normal RNA processing producing a degenerative syndrome.

V. FUTURE DIRECTIONS

In an era of remarkable progress in molecular biology, of cloning and sequencing, it is incongruous that there exists a pathogen which we are not certain contains nucleic acid. However, it now seems that scrapie research is about to come of age. Recent findings have identified distinct fibrilar structures, called scrapie-associated fibrils (SAF),[40] which copurify with infectivity.[41] Examination of proteins in protease-treated preparations of SAF,[37] or similar "rod-like" structures,[42] have detected a 26,000 dalton protein which some investigators are pursuing to determine if it is the scrapie agent, or "prion".[43] Other studies[34,44] have urged a more cautious interpretation of these data, suggesting that the 26,000 dalton protein may be the result of an increased state of aggregation in the scrapie preparation producing differential protection to proteolytic digestion.

It is not clear at this time whether the SAF represent abnormal cytoskeletal elements, a filamentous virus,[45] or ribonucleoprotein matrix fibrils. The latter possibility deserves further examination, especially in light of the finding that SAF preparations[46] contain large amounts of low molecular weight RNA.[34] Previous studies characterizing SAF have focused on protein profiles. Future studies need to be directed at examining these preparations for small nucleic acids. Recent studies on RNA content of membrane vesicles from scrapie-infected hamster brain have shown distinct differences compared to normal brain.[47,48] If a scrapie-specific nucleic acid can be identified, probes can be constructed to test whether it is a component of the transmissible agent, or a product of abnormal processing or tissue damage. The key to understanding this unusual group of animal pathogens, like plant viroids, may be in characterizing individual, disease-specific, small nucleic acids.

REFERENCES

1. **Hanson, R. P. and Marsh, R. F.,** The biology of scrapie and transmissible mink encephalopathy, in *Slow Virus Diseases,* Zeman, W. and Lennette, E. H., Eds., Williams & Wilkins, Baltimore, 1973, 10.
2. **Millson, G. C., Hunter, G. D., and Kimberlin, R. H.,** An experimental examination of the scrapie agent in cell membrane mixtures. II. Association of scrapie activity with membrane mixtures, *J. Comp. Pathol.,* 81, 255, 1971.
3. **Semancik, J. S., Marsh, R. F., Geelen, J. L., and Hanson, R. P.,** Properties of the scrapie agent-endomembrane complex from hamster brain, *J. Virol.,* 18, 693, 1976.
4. **Marsh, R. F., Castle, B. E., Dees, C., and Wade, W. F.,** Equilibrium density gradient centrifugation of the scrapie agent in Nycodenz, *J. Gen. Virol.,* 65, 1963, 1984.
5. **Millson, G. C., Hunter, G. D., and Kimberlin, R. H.,** The physicochemical nature of the scrapie agent, in *Slow Virus Diseases of Man and Animals,* Kimberlin, R. H., Ed., North Holland, Amsterdam, 1976, 243.
6. **Eklund, C. M., Hadlow, W. J., and Kennedy, R. C.,** Some properties of the scrapie agent and its behavior in mice, *Proc. Soc. Exp. Biol. Med.,* 112, 974, 1963.
7. **Malone, T. G., Marsh, R. F., Hanson, R. P., and Semancik, J. S.,** Evidence for the low molecular weight nature of scrapie agent, *Nature (London),* 278, 575, 1979.
8. **Prusiner, S. B., Groth, D. F., Bilstein, C., Masiarz, F. R., McKinley, M. P., and Cochran, S. P.,** Electrophoretic properties of the scrapie agent in agarose gels, *Proc. Natl. Acad. Sci. U.S.A.,* 2984, 1980.

9. **Prusiner, S. B., Hadlow, W. J., Eklund, C. M., and Race, R. E.,** Sedimentation properties of the scrapie agent, *Proc. Natl. Acad. Sci. U.S.A.*, 4656, 1977.
10. **Alper, T., Cramp, W. A., Haig, D. A., and Clarke, M. C.,** Does the scrapie agent replicate without a nucleic acid?, *Nature (London),* 214, 764, 1967.
11. **Latarjet, R., Muel, B., Haig, D. A., Clarke, M. C., and Alper, T.,** Inactivation of the scrapie agent by near monochromatic ultraviolet light, *Nature (London),* 227, 1341, 1970.
12. **Latarjet, R.,** Inactivation of the agents of scrapie, Creutzfeldt-Jakob disease and kuru by radiations, in *Slow Transmissible Diseases of the Nervous System,* Vol. 2, Prusiner, S. B. and Hadlow, W. J., Eds., Academic Press, New York, 1979, 397.
13. **McKinley, M. P., Masiarz, F. R., Isaacs, S. T., Hearst, J. E., and Prusiner, S. B.,** Resistance of the scrapie agent to inactivation by psoralens, *Photochem. Photobiol.*, 37, 539, 1983.
14. **Marsh, R. F., Semancik, J. S., Medappa, K. C., Hanson, R. P., and Rueckert, R. R.,** Scrapie and transmissible mink encephalopathy: search for infectious nucleic acid, *J. Virol.*, 13, 993, 1974.
15. **Cho, H. J.,** Inactivation of the scrapie agent by pronase, *Can. J. Comp. Med.*, 47, 494, 1983.
16. **Prusiner, S. B., McKinley, M. P., Groth, D. F., Bowman, K. A., Mock, N. J., Cochran, S. P., and Masiarz, F. R.,** The scrapie agent contains a hydrophobic protein, *Proc. Natl. Acad. Sci. U.S.A.*, 78, 6675, 1981.
17. **Lax, A. J., Millson, G. C., and Manning, E. J.,** Involvement of protein in scrapie agent infectivity, *Res. Vet. Sci.*, 34, 155, 1983.
18. **Manuelidis, L. and Manuelidis, E. E.,** Search for specific DNA in Creutzfeldt-Jakob infectious brain fractions using nick translation, *Virology,* 109, 435, 1981.
19. **Prusiner, S. B., McKinley, M. P., Bowman, K. A., Bolton, D. C., Bendheim, P. E., Groth, D. F., Braunfeld, M. B., and Barry, R. A.,** Scrapie prions, amyloid plaques and Alzheimer's disease, *Clin. Res.*, 32, 515, 1984.
20. **Prusiner, S. B.,** Novel proteinaceous infectious particles cause scrapie, *Science,* 216, 146, 1982.
21. **Diener, T. O.,** Is the scrapie agent a viroid?, *Nature (London),* 235, 218, 1972.
22. **Ward, R. L., Porter, D. D., and Stevens, J. G.,** Nature of the scrapie agent: evidence against a viroid, *J. Virol.*, 14, 1099, 1974.
23. **Dees, C., German, T. L., Wade, W. F., and Marsh, R. F.,** Characterization of lipids in membrane vesicles from scrapie-infected hamster brain, *J. Gen. Virol.*, 66, 861, 1985.
24. **Lerner, M. J. and Steitz, J. A.,** Snurps and scyrps, *Cell,* 25, 298, 1981.
25. **Wagner, R. R., McGown, J. J., and Grinnell, B. W.,** Pathogenicity of vesicular stomatitis virus is potentially mediated by wild-type leader RNA which inhibits initiation of transcription, in *Mechanisms of Viral Pathogenesis,* Kohn, A. and Fuchs, P., Eds., Matinus Nijhoff, Boston, 1983, 164.
26. **Gallinaro, H., Gattoni, R., Stevenin, J., and Jacob, M.,** Small nuclear RNA and VA RNA in nuclear ribonucleoprotein fibrils from adenovirus-2 infected HeLa cells, *Biochem. Biophys. Res. Commun.*, 95, 20, 1980.
27. **Wolin, S. L. and Steitz, J. A.,** The Ro small cytoplasmic ribonucleoproteins: identification of the antigenic protein and its binding site on the Ro RNA, *Proc. Natl. Acad. Sci. U.S.A.*, 81, 1996, 1984.
28. **Calvet, J. P. and Pederson, T.,** Base-pairing interactions between small nuclear RNAs and nuclear RNA precursors as revealed by psoralen cross-linking in vivo, *Cell,* 26, 363, 1981.
29. **Kahn, G. and Davis, B. P.,** In vitro studies on longwave ultraviolet light dependent reactions of the skin photosensitizer chlorpromazine with nucleic acids, purines and pyrimidines, *J. Invest. Dermatol.*, 55, 47, 1970.
30. **Ogido, T., Iwaki, M., and Mori, K.,** Fluidity of human erythrocyte membrane and effect of chlorpromazine on fluidity and phase-separation of membrane, *Biochem. Biophys. Acta,* 649, 325, 1981.
31. **Dees, C., Wade, W. F., German, T. L., and Marsh, R. F.,** Inactivation of the scrapie agent by ultraviolet irradiation in the presence of chlorpromazine, *J. Gen. Virol.*, 66, 845, 1985.
32. **Prusiner, S. B., Cochran, S. P., Groth, D. F., Bowman, K. A., and Martinez, H. M.,** Measurement of the scrapie agent using an incubation time interval assay, *Ann. Neurol.*, 11, 353, 1981.
33. **Prusiner, S. B., McKinley, M. P., Bowman, K. A., Bolton, D. C., Bendheim, P. E., Groth, D. F., and Glenner, G. G.,** Scrapie prions aggregate to form amyloid-like birefringent rods, *Cell,* 35, 349, 1983.
34. **Castle, B. E., Dees, C., German, T. L., and Marsh, R. F.,** Effects of different methods of purification on aggregation of scrapie infectivity, *J. Gen. Virol.*, in press.
35. **Walter, P., and Blobel, G.,** Signal recognition particle contains a 7S RNA essential for protein translocation across the endoplasmic reticulum, *Nature (London),* 299, 691, 1982.
36. **Branch, A. D. and Robertson, H. D.,** A replication cycle for viroids and other small infectious RNAs, *Science,* 223, 450, 1984.
37. **Buyukmihci, N., Goehring-Harmon, F., and Marsh, R. F.,** Neural pathogenesis of scrapie after intraocular inoculation of hamsters, *Exp. Neurol.*, 81, 396, 1983.
38. **Kuilla, M. G. and Keene, J. D.,** The leader of vesicular stomatitis virus is bound by a cellular protein reactive with anti-La lupus antibodies, *Cell,* 34, 837, 1983.

39. **Van Eekelenc, A. G., Mariman, E. C., Reinders, R. J., and Van Venrooij, W. J.,** Adenoviral heterogeneous nuclear RNA is associated with host cell proteins, *Eur. J. Biochem.*, 119, 461, 1981.
40. **Merz, P. A., Somerville, R. A., Wisniewski, H. M., and Igbal, K.,** Abnormal fibrils from scrapie-infected brain, *Acta Neuropathol.*, 54, 63, 1981.
41. **Diringer, H., Gelderblom, H., Hilmert, H., Ozel, M., Edelbluth, C., and Kimberlin, R. H.,** Scrapie infectivity, fibrils, and low molecular weight protein, *Nature (London)*, 306, 476, 1983.
42. **Bolton, D. C., McKinley, M. P., and Prusiner, S. B.,** Identification of a protein that purifies with the scrapie prion, *Science*, 218, 1309, 1982.
43. **Prusiner, S. B.,** Prions, *Sci. Am.*, 251, 50, 1984.
44. **Dees, C., German, T. L., Wade, W. F., and Marsh, R. F.,** Characterization of proteins in membrane vesicles from scrapie-infected hamster brain, *J. Gen. Virol.*, 66, 861, 1985.
45. **Merz, P. A., Rohwer, R. G., Kascsak, R., Wisniewski, H. M., Somerville, R. A., Gibbs, C. J., Jr., and Gajdusek, D. C.,** Infection-specific particle from the unconventional slow virus diseases, *Science*, 225, 437, 1984.
46. **Hilmert, H. and Diringer, H.,** A rapid and efficient method to enrich SAF-protein from scrapie brains of hamsters, *Biosci. Rep.*, 4, 165, 1984.
47. **German, T. L., McMillan, B., Castle, B. E., Dees, C., Wade, W. F., and Marsh, R. F.,** Comparison of RNA from healthy and scrapie-infected hamster brain, *J. Gen. Virol.*, 66, 839, 1985.
48. **Dees, C., McMillan, B. C., Wade, W. F., German, T. L., and Marsh, R. F.,** Characterization of nucleic acids in membrane vesicles from scrapie-infected hamster brain, *J. Virol.*, 55, 126, 1985.

INDEX

A

B

C

D

I

K

L

R

S

T

U

V

W

X

Y

Z